Tumorimmunologie und Tumortherapie

Beiträge zur Onkologie
Contributions to Oncology

Band 25

Reihenherausgeber
S. Eckhardt, Budapest; *J. H. Holzner,* Wien;
G. A. Nagel, Göttingen

Basel · München · Paris · London · New York · New Delhi · Singapore · Tokyo · Sydney

Tumorimmunologie und Tumortherapie

Eine Standortbestimmung

H. H. Sedlacek, Marburg

2 Abbildungen (in Farbe) und 6 Tabellen, 1987

KARGER

Basel · München · Paris · London · New York · New Delhi · Singapore · Tokyo · Sydney

Beiträge zur Onkologie
Contributions to Oncology

CIP-Kurztitelaufnahme der Deutschen Bibliothek

Sedlacek, Hans Harald:
Tumorimmunologie und Tumortherapie: e. Standortbestimmung / Hans H. Sedlacek. –
Basel; München; Paris; London; New York; New Delhi; Singapore; Tokyo; Sydney:
Karger, 1987.
ISBN 3-8055-4447-2
NE: GT
(Beiträge zur Onkologie; Vol. 25)

Dosierungsangaben von Medikamenten

    Autoren und Herausgeber haben alle Anstrengungen unternommen, um sicherzustellen, daß Auswahl und Dosierungsangaben von Medikamenten im vorliegenden Text mit den aktuellen Vorschriften und der Praxis übereinstimmen. Trotzdem muß der Leser im Hinblick auf den Stand der Forschung, Änderungen staatlicher Gesetzgebungen und den ununterbrochenen Strom neuer Forschungsergebnisse bezüglich Medikamenteneinwirkung und Nebenwirkungen darauf aufmerksam gemacht werden, daß unbedingt bei jedem Medikament der Packungsprospekt konsultiert werden muß, um mögliche Änderungen im Hinblick auf die Indikation und Dosis nicht zu übersehen. Gleiches gilt für spezielle Warnungen und Vorsichtsmaßnahmen. Ganz besonders gilt dieser Hinweis für empfohlene neue und/oder nur selten gebrauchte Wirkstoffe.

Alle Rechte vorbehalten.

    Ohne schriftliche Genehmigung des Verlags dürfen diese Publikation oder Teile daraus nicht in andere Sprachen übersetzt oder in irgendeiner Form mit mechanischen oder elektronischen Mitteln (einschließlich Fotokopie, Tonaufnahme und Mikrokopie) reproduziert oder auf einem Datenträger oder einem Computersystem gespeichert werden.

©   Copyright 1987 by S. Karger GmbH, Postfach, D-8034 Germering/München und
S. Karger AG, Postfach, CH-4009 Basel
Printed in Germany by Gebr. Parcus KG, München
ISBN 3-8055-4447-2

Inhalt

Vorwort VII

Geleitwort VIII

Verzeichnis der verwendeten Abkürzungen IX

Einleitung 1

Tumorantigene 8
    Nachweismethoden 8
        Nachweis über Transplantationsversuche............ 8
        Nachweis über zellvermittelte Reaktionen 10
        Nachweis mit Hilfe von Antikörperreaktionen........ 15
    Klassifizierung der Tumorantigene 17
    Tumorspezifische Oberflächen («Surface»)-Antigene (TSSA)... 17
    Tumorassoziierte Antigene (TAA) 18
    Embryonale Antigene 21
    Organ- bzw. gewebespezifische Antigene............. 21
    Kryptische Antigene 22
    Tumorantigene und Histokompatibilitätsantigene 23

Tumor und Immunsystem................................... 25
    Antigenspezifische Immunabwehr 26
        Erkennung von Antigenen, inklusiv von Tumorantigenen (afferent) 26

        Zelluläre Abwehrreaktionen (zytotoxische T-Zellen
        und Reaktionen vom verzögerten Typ) 30
        Humorale Abwehrreaktionen 34
Unspezifische Immunabwehr 35
        Makrophagen 35
        Natürliche Killer-(NK-)Zellen 42
        Granulozyten 47
Antikörperabhängige, zelluläre Zytotoxizität (Antibody-
dependent cellular cytotoxicity = ADCC) 47
Beeinträchtigung der Immunabwehr 49
        Quantitative oder qualitative Mängel des Immunsystems 50
        Mangelnde Tumorerkennung 51
        Abscheidung von Tumorantigenen 53
        Inhibition von Makrophagen 53
        Aktivierung von Suppressorzellen 55
        Hemmung durch Antikörper 57
        Hemmung durch Antigen-Antikörperkomplexe 57

Ansatzpunkte zur Tumorimmuntherapie 59
Passive Immuntherapie mit Antikörpern 59
Passive Immuntherapie mit Lymphozyten und Produkten
von Lymphozyten 64
Aktive unspezifische Immunmodulation 68
Aktive spezifische Immuntherapie 71
        Inaktivierte Tumorzellen 72
        Zellhomogenate 74
        Virusonkolysate 75
        Chemisch behandelte Tumorzellen 75
        Mit Enzymen behandelte Tumorzellen 77

Zusammenfassung und Wertung 85

Danksagung 90

Literatur 91

Vorwort

Mit zunehmendem Einblick in die Regelvorgänge des Immunsystems sind auch die Hoffnungen gewachsen, die Abwehrmechanismen des Immunsystems für eine Therapie von Tumoren nutzen zu können.

So ist bereits seit längerem versucht worden, Antigen-unspezifisch mit sogenannten Immunstimulanzien das Abwehrsystem von Tumorpatienten zu steigern oder aber durch Verabreichung von Tumorassoziierten Antigenen im Gemisch mit geeigneten Adjuvanzien eine aktive spezifische Tumortherapie durchzuführen. Neuere Bestrebungen laufen darauf hinaus, mit monoklonalen Antikörpern oder mit Wirkstoffen (Mediatoren) des Immunsystems eine Tumortherapie zu erzielen. Klinische Untersuchungen im größeren Ausmaß sind möglich geworden durch die großtechnische Herstellung dieser Substanzen mit Hilfe von zellbiologischen und molekularbiologischen Techniken. Die Wirksamkeitsprüfung von Tumorimmuntherapeutika an Patienten ist teils bereits abgeschlossen, teils in vollem Gange. Enttäuschende Ergebnisse, wie beispielsweise mit Immunstimulanzien, aber auch hoffnungsvolle Resultate wurden erzielt. Zwar sind wir noch weit davon entfernt, die Immuntherapie als ein etabliertes Verfahren zur Therapie von Tumoren ansehen zu dürfen, aber es bestehen begründete Aussichten, daß zukünftig auch immunologisch-therapeutische Verfahren teilhaben werden an dem «scheibchenweise» erzielten Erfolg im Kampf gegen Tumoren.

Marburg, Juli 1986 *H. H. Sedlacek*

Geleitwort

«Stahl und Strahl» in der Behandlung von Tumorerkrankungen sind erfolgreich durch die Chemotherapie ergänzt worden. Doch trotz dieser Erfolge stehen wir in der Onkologie noch am Anfang: Nur ca. 7 bis 10% disseminierter Tumorerkrankungen können geheilt werden, ca. 50% trotzen jeder Therapie, ca. 40% sind mit den heutigen therapeutischen Methoden nur vorübergehend beeinflußbar. Diese insgesamt enttäuschende Bilanz sollte den Mut wecken, neue Wege in der Therapie von Tumoren zu bahnen: Das Handwerkszeug hierfür ist uns durch die lawinenartige Wissenszunahme auf dem Gebiet der Molekularbiologie, der Zellkulturtechnik und der Immunologie gegeben worden. Doch aller neuer Anfang ist schwer und langwierig − so auch in der Tumorimmuntherapie. Eine kritische Bilanz des fast ausufernden Wissens auf diesem Gebiet ist mit der vorliegenden Arbeit gezogen worden. Manche therapeutischen Wege haben sich als Sackgasse erwiesen. Andere konnten Hoffnungen wecken. So bestehen begründete Aussichten, daß dem Onkologen mit monoklonalen Antikörpern, mit ausgewählten Immunmediatoren oder durch «Immunisierung» mit Tumorzellantigenen im Gemisch mit Neuraminidase in absehbarer Zeit bei bestimmten Tumoren eine zusätzliche therapeutische Möglichkeit in die Hand gegeben werden kann. Der Weg dorthin benötigt jedoch neben Zeit, Beharrlichkeit und einer fortwährenden kritischen Überprüfung des jeweiligen Standortes auch die Zusammenarbeit unterschiedlicher, gerade auch klinischer Disziplinen. Mein Wunsch ist es, daß dieses Buch diese Zusammenarbeit fördern möge.

Marburg, Mai 1986 *H. G. Schwick*

Verzeichnis der verwendeten Abkürzungen

ADCC	«Antibody-dependent cellular cytotoxicity»
AML	Akute myeloische Leukämie
APC	Antigen-präsentierende Zellen
asialo GM_1	Asialo-Ganglio-tetraosyl-ceramid
ATLA	«Adult T-cell leukemia antigen»
ATLV	«Adult T-cell leukemia virus»
BCG	Bacillus Calmette-Guérin
B-Zellen	«Bone-marrow»-differenzierte Lymphozyten
CEA	«Carzinoembryonales» Antigen
CFA	«Complettes» Freundsches Adjuvans
CI	Colony inhibition
CLL	Chronisch lymphatische Leukämie
CMC	Cell-mediated cytotoxicity
C. parvum	Corynebakterium-parvum-Präparation
C.P.	C. parvum
ConA	Concavalin A
CSF	«Colony»-stimulierender Faktor
CTL	Zytotoxische T-Lymphozyten
DNA/DNS	Desoxyribonukleinsäure
DTH	Delayed type hypersensitivity
EA	«Early antigen»
EBV	Epstein-Barr-Virus
E. coli	Escherichia coli
ELISA	Enzyme-linked immunosorbent assay
Fc	Fragment cristalline von Ig
HA	Histokompatibilitätsantigene
HLA	Humane Leukozyten-Antigene

Verzeichnis der verwendeten Abkürzungen

HTLV	Human T-cell leukemia virus
HvG	Host versus graft
Ia-Antigen	«Immune response» Antigen (der Maus)
IFA	Inkomplettes Freundsches Adjuvans
Ig	Immunglobulin
IR-Gene	«Immune response»-Gene
K	Kappa (Isotyp der leichten Kette des Immunglobulins)
LAI	Leukozyten-Adhärenz-Inhibition
LGL	Large granular lymphocytes
LMI	Leukozyten-Migration-Inhibition
LPO	Laktoperoxydase
L-Kette	Leichte Kette (des Immunglobulins)
LAF	Lymphozyten-aktivierender Faktor (Interleukin I)
MAF	Makrophagen-aktivierender Faktor
MAk	Monoklonale Antikörper
MDP	Muramyldipeptid
MHA	«Mixed hemabsorption assay»
MHC	«Major histocompatibility complex»
MER	Methanol extracted residue von BCG
MIF	Migrations-Inhibitions-Faktor
MLC	«Mixed lymphocyte culture»-Technik
MMI	Makrophagen-Migration-Inhibition
MSF	Makrophagen «slowing» Faktor
MW	Molekulargewicht
NCI	National Cancer Institute
NK-Zellen	Natürliche Killer-Zellen
PA	Protein A-Assay
PBS	Phosphat-buffered saline
PFC	Plaque-forming cells
PHA	Phytohämagglutinin
RNA	Ribonukleinsäure
RPMI	Rosewell-Park Memorial Institute
TAA	Tumorassoziierte Antigene
T-Antigen	Thomsen-Friedenreich-Antigen
TSSA	Tumorspezifische «Surface»-Antigene
TSTA	Tumorspezifische Transplantationsantigene
T-Zelle	Thymus-geprägte Lymphozyten
VCN	Vibrio cholerae Neuraminidase
WHO	World Health Organization

Einleitung

Die von Burnet [190, 191] und Thomas [1494] entwickelte Hypothese der «Immune surveillance» besagt, daß entartete Zellen durch das Immunsystem des Wirtes erkannt und eliminiert werden. Demzufolge sollte die Entstehung eines Tumors im wesentlichen durch eine primäre oder sekundäre Insuffizienz des Immunsystems verursacht sein.

Die Theorie von Burnet fußt auf zahlreichen, seit der Mitte des vergangenen Jahrhunderts durchgeführten klinisch empirischen, meist kasuistischen Studien zur Immuntherapie von Tumoren (siehe Tabelle I). Sie hat die heutigen Vorstellungen über Wechselwirkungen zwi-

Tabelle I. Geschichtliche Entwicklung der klinischen Tumorimmuntherapie

Jahr	Autor/Referenznummer	
Antigen-spezifisch aktiv		
1902	v. Leiden und Blumenthal [1576]	Tumorzellen (autolog)
1909	v. Dungern [1575]	Tumorzellbrei
1911	Risley [1207]	Extrakt aus Tumorgewebe (autolog oder allogen)
1913	Pinkuss [1132]	Phenol-behandelte Tumorzellen
1922	Kellock et al. [757]	Tumorfragmente (autolog)
1960	Finney et al. [404]	Tumorhomogenat in CFA
1967	Czajkowski et al. [266]	Tumorzellen, an die chemisch Kaninchenantikörper gekoppelt wurden
1969	Currie und Bagshawe [259]	Neuraminidase-behandelte Tumorzellen
1970	Hughes et al. [670]	Tumorextrakt, vermischt mit Bordetella pertussis vaccine oder CFA
1971	Currie et al. [260]	bestrahlte Tumorzellen

Tabelle I. (Fortsetzung)

Antigen-spezifisch passiv

1895	Hericourt und Richet [612]	Antiseren von Hunden und Eseln (immunisiert mit Tumormaterial)
1901	Boeri [122]	Antiseren von der Ziege (immunisiert mit Tumormaterial)
1958	Murray [1015]	Antikörper vom Pferd (immunisiert mit Tumormaterial)
1959	Buinauskas et al. [181]	Antikörper vom Schaf (immunisiert mit Tumormaterial)
1960	Sumner und Foraker [1459]	Vollblut von Patienten mit Tumorregression
1968	Laszlo et al. [857]	Isoantikörper gegen Lymphozyten (Therapie der CLL)

Antigen-unspezifisch aktiv

1868	Busch [192]	Streptokokkeninfektionen bewirken Tumornekrose
1893	Coley [237]	bakterielle Toxine (Serratia marcescens)
1938	Fedyushin [379]	Antikörper gegen Komponenten des retikuloendothelialen Systems
1968/1969	Mathé et al. [932, 933]	Bacillus Calmette-Guérin (BCG)
1974	Israel et al. [697]	Tilorone (basischer Äther von Fluorenon)
1975	Halpern [543]	Corynebacterium parvum (Formalin-behandelt)
1975	Yamamura et al. [1630]	«Cell wall skeleton»-Substanz (Mycolsäure-Arabinogalactan-Mycopeptid) von BCG
1976	Weiss [1598]	Methanol-extracted residue (MER) von Phenol-behandeltem BCG
1978	Azuma et al. [40]	«Cell wall skeleton»-Substanz von Nocardin rubra
1978	Gee et al. [455]	Lipopolysaccharid aus Pseudomonas aeruginosa
1978	Bicker [106]	
1981	Goutner et al. [502]	Azimexon (2-Cyanazividin-Derivat)
1981	Uchida und Hoshino [1532]	β-hämolytische Streptokokken, Penicillin-behandelt
1981	Serrou et al. [1350]	Bestatin, ein Oligopeptid und Inhibitor von Aminopeptidase Levamisole (Tetramisole)

Tabelle I. (Fortsetzung)

Antigen-spezifisch adoptiv

1969	Nadler und Moore [1018]	Blutlymphozyten von Spendern, immunisiert mit Tumoren
1967	Andrews et al. [24]	allogene Ductus-thoracicus-Lymphozyten von Spendern, immunisiert mit Tumoren
1971	Humphrey et al. [674, 675]	Serum und Blutleukozyten von Spendern, immunisiert mit Tumoren
1967	Alexander et al. [13]	Nukleinsäure aus lymphoiden Zellen von spezifisch immunisierten Spendern (syngen, xenogen)

Antigen-unspezifisch adoptiv

1963	Woodruff und Nolan [1618]	Milzzellen eines gesunden Spenders
1965	Mathé [928]	allogene Knochenmarktransplantation (Leukämie)
1966	Schwarzenberg et al. [1307]	Infusion von Leukozyten (CLL) in Patienten mit akuter Leukämie
1968	Symes et al. [1466]	allogene und xenogene Lymphozyten

Lokal

1968	Klein [791–793]	Auslösung einer Immunreaktion vom verzögerten Typ mit Hilfe von DNCB am Tumor
1969		
1970	Morton et al. [1004]	intratumorale Injektion von BCG
1971	Hunter-Craig et al. [676]	intratumorale Injektion von Vaccinia

schen Tumor und Immunsystem entscheidend geprägt und ist Veranlassung zu einer großen Zahl experimenteller und klinischer Untersuchungen gewesen.

So wurde in den letzten zwei Jahrzehnten intensiv versucht, das Wechselspiel zwischen Immunsystem und Tumor zu erfassen und hier vorhandene Abwehrmechanismen aufzuspüren. Des weiteren wurden, experimentell und auch klinisch, verschiedene Möglichkeiten der Immuntherapie des Krebses (siehe hierzu Tabelle II) erprobt, die «spezifische Immuntherapie» durch Verabreichung einer «Vaccine», bestehend aus Tumorzellen (oder hieraus hergestellten Antigenen) und einem Adjuvans, oder aber die «unspezifische Immunmodulation» mit Hilfe von

Tabelle II. Möglichkeiten der Tumorimmuntherapie

	Passiv	Aktiv
Antigen-unspezifisch	*Mediatoren* Interferone Interleukine Lymphotoxine Tumornekrosisfaktoren koloniestimulierende Faktoren (CSF) *Komplementfaktoren* *K-Zellen; NK-Zellen*	*Immunstimulanzien* Mikroorganismen und mikrobielle Produkte synthetische Chemoimmuntherapeutika

Antigen-spezifisch	*Antikörper*
	
	gegen Tumoren
	– IgG$_{2a}$; IgG$_3$ (Maus)
	– radioaktiv markierte
	  Antikörper
	– Immunzytostatika;
	  Immuntoxine
	
	Lymphozyten
	
	– spezifisch sensibilisierte
	  T-Zellen
		Tumorzellen + Adjuvans
		
		Tumorzellen + Neuraminidase
		Tumorzellen + BCG
		Tumorantigene + Adjuvans
		
		*Hybride aus Tumorzellen und
		Normalzellen*
		
		Virus-infizierte Tumorzellen

bakteriellen Inhaltsstoffen oder synthetischer, durch «Screening» entdeckter neuer Substanzen.

Der klinische Erfolg, den man in geeignet kontrollierten prospektiven Studien zu ermitteln versuchte, ist jedoch in den meisten Fällen ausgeblieben (Übersicht bei [1487]). Auch andere Formen der Immuntherapie, wie beispielsweise die Verabreichung von in vitro spezifisch

---

Abb. 1. Immunologische Tumorabwehr.
▲ Tumorassoziiertes Antigen;
● Histokompatibilitäts-Antigene (MHC-Klasse II);
Ⲩ Antikörper;
□■ «Rezeptoren» für natürliche Killer-Zellen und Makrophagen.

Einleitung

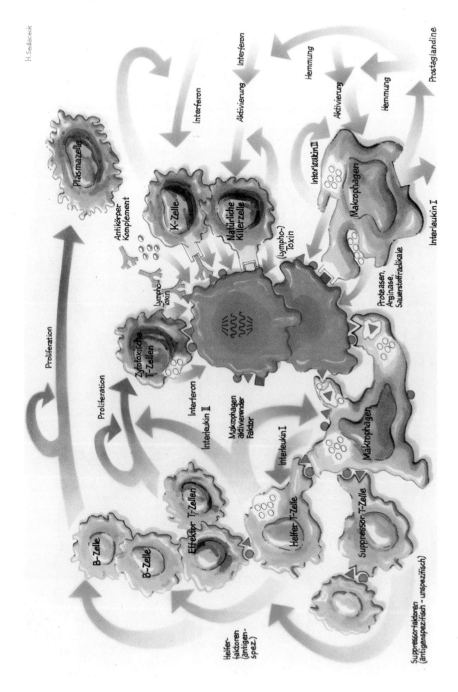

Tumorimmunologie und Tumortherapie

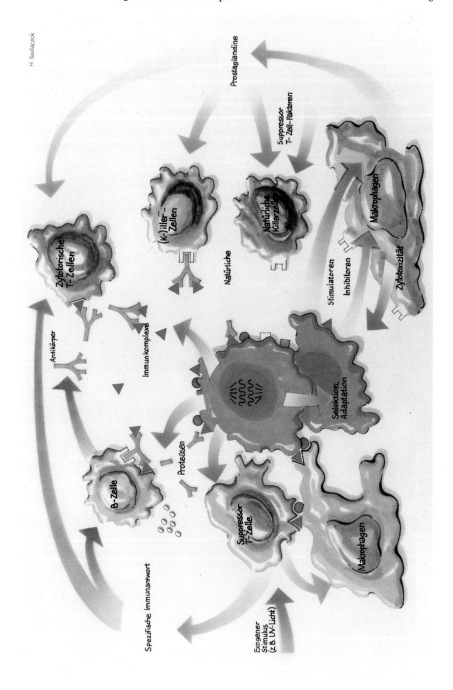

sensibilisierten Lymphozyten, von Serumfaktoren wie Antikörper und Komplement oder mehr oder weniger gereinigten bzw. über gentechnologische Methoden hergestellten Leukozytenmediatoren haben bislang nicht den ersehnten Durchbruch gebracht.

Diese ernüchternde Bilanz macht den derzeitigen grundsätzlichen Zweifel an der Chance der Immuntherapie als Behandlungsmöglichkeit des Krebses verständlich. Trotz all dieser Zweifel dürfen jedoch nicht solche Ansätze zur Tumorimmuntherapie übersehen werden, welche sowohl in experimentellen als auch klinischen Untersuchungen Ergebnisse geliefert haben, die wegweisend sein könnten für eine weitere Entwicklung der Tumorimmuntherapie hin zu einer wirksamen therapeutischen Waffe. Zu diesen aus heutiger Sicht zumindest hoffnungsvollen Ansätzen darf im besonderen die aktive, antigenspezifische Tumortherapie mit Neuraminidase und Tumorzellen zum einen und die passive antigenspezifische Tumortherapie mit monoklonalen Antikörpern zum anderen gezählt werden.

Abb. 2. Beeinträchtigung der Immunabwehr durch Tumorzellen. (Zeichenerklärung siehe Abbildung 1).

Tumorantigene

Die Kontrolle des Tumorwachstums durch das Immunsystem hat zur Voraussetzung, daß sich Tumorzellen von Normalzellen durch qualitative oder quantitative Merkmale auf der Zellmembran unterscheiden und somit von dem Immunsystem erkannt werden können.

Nachweismethoden

Nachweis über Transplantationsversuche

Seit längerem war durch Transplantationsversuche die Existenz von Tumorantigenen vermutet worden [337]. Der Beweis für ihre Existenz konnte jedoch erst angetreten werden, nachdem Transplantationsantigene entdeckt worden waren und von diesen tumorspezifische Transplantationsantigene (TSTA) unterschieden werden konnten [499, 500], des weiteren nachdem genetisch einheitliche Inzuchtstämme von Mäusen zur Verfügung standen, die Transplantationsexperimente in syngenen Systemen ermöglichten. Mit Hilfe derartiger Experimente in einem C 3 H-Inzuchtstamm wurde erstmals gezeigt, daß Mäuse durch Immunisierung mit Material eines durch Methylcholanthren induzierten Tumors gegen das Anwachsen eines nachfolgenden Transplantats des gleichen Tumors zu schützen sind [522]. Nachfolgende Untersuchungen ergaben, daß auch ein wachsendes Transplantat eines chemisch induzierten Tumors in der Maus eine Immunreaktion erzeugt, die nach Exzision dieses Tumors durch die entstandene Resistenz gegen einen nachfolgend transplantierten zweiten Tumor des gleichen Typs nachgewiesen werden konnte [46, 414, 415].

Die Tumorspezifität dieser Reaktionen wurde durch geeignete Kontrollen nachgewiesen, wie durch parallel durchgeführte Transplantation der Haut des Tumorspenders [1166] oder durch Versuche in autochthonen Systemen [795].

Entsprechend der Nachweistechnik wurden die auf dem Tumor ermittelten Antigene als tumorspezifische Transplantationsantigene (TSTA) bezeichnet. Ihr Vorkommen wurde sowohl auf chemisch und physikochemisch induzierten als auch auf Virus-induzierten Tumoren [1383–1386] und auf sogenannten spontanen, d. h. ohne erkennbare Ursachen entstandenen, experimentellen Tumoren untersucht [201, 1548]. Während die TSTA von chemisch induzierten Tumoren eine extreme Heterogenität aufweisen, so daß bei verschiedenen in einem Tier induzierten Tumoren meist keine immunologische Kreuzreaktivität nachzuweisen ist [71, 523, 795, 796, 1074] und sogar innerhalb eines Tumors extrem unterschiedliche, unabhängige TSTA auftreten können [1623], zeichnen sich zahlreiche, durch Viren experimentell oder natürlich induzierte Tumoren (Übersicht bei [1191]) dadurch aus, daß sie einheitliche virusspezifische TSTA besitzen (Übersicht bei [847, 1068]). So konnten Sjögren et al. [1386] nach Infektion von Mäusen mit Polyomaviren einen Schutz gegen die Transplantation von Polyomavirus-induzierten Tumoren erzeugen. Dieser Schutz war auch nach Immunisierung mit Polyoma-induzierten Tumorzellen [1384, 1385] zu erreichen, sowohl in Mäusen als auch in Hamstern [528]. Die Entwicklung einer Transplantationsimmunität gegen diese Tumoren konnte durch Bestrahlung der Empfängertiere [1383] beeinträchtigt oder auf unbehandelte Tiere durch Lymphozyten von spezifisch immunisierten Tieren übertragen werden [286, 529]. Ähnliche Ergebnisse wurden auch bei anderen Virus-induzierten Tumoren beobachtet (z. B. bei SV40-induzierten Tumoren [283, 530, 770] oder bei Skopepapillomen [575, 953]). Zusätzlich zu einheitlichen virusspezifischen TSTA können jedoch, wie beim Mammatumorvirus der Maus nachgewiesen [1003], auch individual-spezifische TSTA auftreten.

Die jeweiligen TSTA können sowohl bei chemisch- als auch bei Virus-induzierten Tumoren in außerordentlich unterschiedlicher Menge ausgeprägt sein. Demgegenüber sind Spontantumoren häufig entweder gering oder nicht immunogen [47, 624, 1163]. Jedoch können derartige Spontantumoren durch Behandlung mit einem Karzinogen TSTA ausbilden, die nicht nur einen Immunschutz für die jeweilige Mutante, sondern auch für die unbehandelte ursprüngliche TSTA-negative Tumorzelle liefern [128, 130, 1557]. Diese müßte demnach TSTA ausbilden können, die zwar nicht immunogen, jedoch antigen sind.

Speziell bei chemisch induzierten Tumoren ist ungeklärt, ob die Ausbildung von TSTA ein phänotypischer Marker für neoplastische Transformation ist (Übersicht bei [1111]). Zwar konnte mit Hilfe von DNA-Transfektionsversuchen in 3 T 3-Fibroblasten ein Cotransfer der transformierenden Aktivität und der TSTA-Expression nachgewiesen werden [658], jedoch spricht die extreme Vielfalt von TSTA gegen die Annahme, daß von außen in die Zelle hineingebrachtes genetisches Material ursächlich an der Ausbildung karzinogen induzierter TSTA beteiligt sein könnte.

Derzeit wird diskutiert, ob entweder eine Alteration der Gensysteme, die die Individualspezifität kodieren (wie «major and minor histocompatibility complex») oder aber eine Beeinflussung der Gene, welche die Verschiedenheit in einem Individuum bestimmen (wie z. B. die Gene für die Ig-Synthese und für den T-Zell Rezeptor), oder eine Aktivierung von endogenen virusverwandten Genstrukturen direkt oder indirekt durch Karzinogene stattfindet (Übersicht bei [1111]). Insgesamt ist es bislang jedoch unbekannt, ob die Ausbildung von TSTA eine charakteristische Eigenschaft von Tumoren darstellt und ob methodische Unzulänglichkeiten der alleinige Grund für ihre mangelnde Nachweisbarkeit auf einer Großzahl von Tumorzellen sind [1068]. Um die Empfindlichkeit der Nachweismethoden zu erhöhen und auch um Tumorantigene des Menschen erfassen zu können, wurden demzufolge zellvermittelte und serologische Testmethoden aufgebaut.

Nachweis über zellvermittelte Reaktionen

Der Befund, daß eine Immunität gegen TSTA chemisch induzierter Sarkome durch die Übertragung von Lymphknotenzellen oder Milzzellen, nicht aber durch Immunserum, auf nicht-immune, syngene Tiere zu übertragen war [795], legte aufgrund paralleler Erfahrungen in der Organ-Transplantationsimmunologie [582, 1068] nahe, daß die Immunreaktionen gegen TSTA der Reaktion vom verzögerten Typ (Delayed type hypersensitivity reaction = DTH) zuzuordnen sind. Untersuchungen an der Maus [1100] und am Meerschweinchen [1067] bestätigten, daß TSTA in Tieren, die mit dem gleichen Tumorantigen immunisiert worden waren, in der Haut eine DTH auslösen konnte, und daß derartig immunisierte und reaktive Tiere gegen die Transplantation des jeweiligen Tumors spezifisch geschützt waren. Die DTH-Reaktion wurde jedoch negativ, wenn der Tumor im Testtier zu groß wurde [1067]. Sowohl die Immunität gegen den Tumor als auch die verstärkte DTH-Reaktion gegen Tumorantigene war durch Transplantation von Milzzellen zu übertragen [588], die DTH jedoch auch mit Serum, welches Antikörper möglicherweise zytophil für Makrophagen gegen das jeweilige Antigen enthält [588, 671, 1652]. Im Gegensatz hierzu konnte jedoch in anderen Untersuchungen keine Korrelation zwischen der spezifischen Immunität gegen Tumoren und der DTH gegen das jeweilige Tumorantigen gefunden werden [588], was die (tumor-)diagnostische Wertigkeit der DTH-Reaktion in Frage stellt.

Auch bei menschlichen Tumorpatienten war mit Hilfe autologen Tumormaterials eine DTH-Reaktion in der Haut auszulösen, so beim Melanom, beim Burkitt's Lymphom, bei Leukämien, Fibrosarkomen und bei Karzinomen der Mamma, des Kolon, der Lunge, des Magens und der Ovarien (Übersicht siehe [1068]). In einigen Fällen korrelierten diese Werte mit dem klinischen Verlauf der Erkrankung, beispielsweise wiesen Patienten mit lokal begrenzten Tumoren eine stärkere DTH-Reaktion auf als solche mit disseminierter und fortgeschrittener Erkrankung [125, 188, 557]. Es war naheliegend, aus diesen Befunden diagnostische und prognostische Möglichkeiten zu folgern, jedoch wird deren Wert durch die bislang mangelhaften Präparationsmöglichkeiten von Tumorantigenen, die mangelnde Empfindlichkeit der DTH-Reaktion und durch die meist ungelöste Frage nach den geeigneten positiven und negativen Kontrollen eingeschränkt [1068]. Ob im Falle des Thomsen-Friedenreich-Antigens (Übersicht bei [1533]), welches tumorassoziiert sein soll und gegen welches positive DTH-Reaktionen beim Mamma-, Lungen- und Pankreaskarzinom [1430] gefunden wurden, ein Durchbruch in der klinischen Verwertbarkeit der DTH-Reaktion zur Tumordiagnostik erzielt werden kann, ist bislang fraglich.

Angesichts der Nachteile (mangelnde Empfindlichkeit, Notwendigkeit der intradermalen Injektion) der DTH-Reaktion wurde parallel zur allgemeinen Entwicklung immunologischer In-vitro-Techniken auch versucht, mit verschiedenen zellulären Methoden Tumorantigene zu definieren [49, 599, 1068].

Im wesentlichen gehören zu diesen Methoden:
— Die Wachstumsinhibition («Colony inhibition» = CI; [573]) oder die derzeit gebräuchlichere Abtötung von Tumorzellen durch Lympho-

zyten des Tumorträgers (Cell-mediated cytotoxicity = CMC) [74, 1471, 1480]. Der zytotoxische Effekt der Lymphozyten wird wahrscheinlich durch Mediatoren (Lymphotoxin?) verursacht, die von sensibilisierten Lymphozyten, unter anderem auch nach Kontakt mit dem spezifischen Antigen, ausgeschüttet werden [506, 592, 593]. Es existieren verschiedene Versionen der CMC-Technik, so beispielsweise Kurzzeit- (3–4 h) und Langzeittechniken.

– Die Inhibition der aktiven Wanderung von Leukozyten oder von mit Makrophagen angereicherten Zellpräparationen des Tumorträgers durch Tumorzellen oder Tumorextrakte (direkter Test) oder die Inhibition der aktiven Wanderung von Leukozyten oder Makrophagen normaler Spender durch Lymphozyten und Tumorzellen bzw. Tumorextrakte des Tumorträgers (indirekter Test) («Leukocyte» oder «Macrophage migration inhibition» = LMI oder MMI [117, 230, 538]). Die Inhibition der aktiven Wanderung von Leukozyten wird durch einen «Migration-inhibition-factor» (MIF) bewirkt, der von sensibilisierten Lymphozyten nach Kontakt mit dem spezifischen Antigen ausgeschieden wird.

– Die Blastenbildung von sensibilisierten Lymphozyten des Tumorträgers durch spezifische Antigene, das heißt durch Tumorzellen oder Tumorzellextrakte, wird herkömmlicherweise durch Ermittlung der DNS-Syntheserate anhand des Einbaus von radioaktiv markiertem Thymidin bestimmt, entsprechend der («One way»-) «mixed lymphocyte culture technique» (MLC) [42].

– Die Inhibition der Adhärenz von Leukozyten («Leukocyte adherence inhibition» LAI) des Tumorträgers an Glas (Zählkammer) [539] oder an Plastikröhrchen [524, 525, 643] erfolgt durch Tumorzellen oder Tumorextrakte. Möglicherweise wird die Veränderung der Adhärenz von Leukozyten (Monozyten) an Glas durch Lymphokine verursacht, welche von sensibilisierten Lymphozyten nach Kontakt mit dem spezifischen Antigen produziert werden; bei der Verwendung von Plastikröhrchen nimmt man jedoch die Bindung von zytophilen Antikörpern an Monozyten als Ursache an [524].

– Ein Test besonderer Art ist der «elektrophoretische Mobilitäts-Test (EMT)». Im ursprünglichen Verfahren wurden Lymphozyten von Krebspatienten mit dem basischen Protein des Myelin (dem sogenannten «enzephalitogenen Faktor») inkubiert und das von den Lymphozyten sekretierte Lymphokin («Macrophage slowing factor» = MSF) durch Bestimmung der Veränderung der elektrophoretischen Beweglichkeit von Peritonealmakrophagen von Meerschweinchen bestimmt [400].

Der Test wurde durch Einsatz von tannierten, mit Sulfosalizylsäure stabilisierten Schafserythrozyten als Indikatorzellen optimiert [1139]. Erste, sowohl experimentelle [560] als auch klinische Untersuchungen zeigten eine relativ gute Korrelation mit Tumorerkrankungen, aber häufig auch falsch-positive Befunde, besonders bei Virusinfektionen und degenerativen neurologischen Erkrankungen [36, 37, 1139, 1140]. Erste Untersuchungen, den sogenannten «enzephalitogenen Faktor» durch Tumorantigenpräparationen (KCL-Extrakte) zu ersetzen, wiesen auf die Möglichkeit einer organspezifischen, möglicherweise tumorantigenspezifischen Diagnose hin [322, 1009, 1011]. Jedoch ist diese Frage bis heute noch ungeklärt.

Eine Bewertung der zellulären Techniken für die Definition von Tumorantigenen ist nur eingeschränkt möglich: Unter dem Eindruck, daß T-Lymphozyten notwendig sind, um eine spezifische Anti-Tumorimmunität zu übertragen [92, 238, 475, 492], wurden insbesondere T-Zellabhängige Methoden (MMI bzw. LMI; CMC) auf ihre Aussagefähigkeit hin untersucht (Übersicht bei [1044, 1068]). Sowohl bei der Maus [538] als auch beim Meerschweinchen [117] und beim Menschen [230] konnten Tumorantigene mit Hilfe der MMI erfaßt werden, wobei in den tierexperimentellen Untersuchungen die Tumorspezifität dieser Reaktion eng mit jener, erfaßt durch Tumortransplantationsuntersuchungen, korrelierte. Auch der Leukozyten (Makrophagen)-Migrations-Inhibitionstest (LMI) zeigte anfänglich eine relativ hohe Spezifität sowohl für autologes Tumorgewebe als auch für allogenes Tumorgewebe und hieraus hergestellte Tumorantigene [22, 230, 727, 863, 944, 1338]. Keine Inhibition konnte mit Normalgewebe oder mit Zellen von Tumoren anderen histologischen Typs beobachtet werden. Jedoch auch hier traten relativ häufig falsch-positive und falsch-negative Reaktionen auf, so daß eine Verwertung des Tests für die Erkennung von Tumorantigenen fraglich erscheint. Zudem ist ungeklärt, ob positive Reaktionen durch tumorspezifische Antigene bedingt sind oder aber durch auch auf Normalzellen vorkommende, auf Tumorzellen jedoch stärker anzutreffende sogenannte tumorassoziierte Antigene (siehe Übersicht bei [159, 1068]).

Ähnlich widersprüchliche Ergebnisse konnten mit dem «Leukocyteadherence-inhibition»-Test erzielt werden [539, 542, 583, 1149]. Auch hier waren die anfänglich guten, auch tierexperimentellen Ergebnisse [1017] nicht in vollem Umfang zu reproduzieren [228, 1346], möglicherweise aufgrund von Modifikationen in der Technik oder aber wegen der Schwierigkeit, Tumorantigene zu isolieren [1068].

Eine gute Spezifität für Tumorantigene, korrelierend mit den Ergebnissen der Tumortransplantationsversuche, konnte in ersten Untersuchungen an chemisch induzierten Tumoren auch für die «Colony inhibition» gefunden werden [573], jedoch war auffallend, daß Tumoren wuchsen, obwohl die jeweiligen Tumorträger aktive Immunzellen besaßen. Dieses galt sowohl für chemisch als auch für Virus-induzierte oder spontane Tumoren [581] und wurde auf die Variabilität der jeweiligen Testmethode und möglicherweise auch auf einen den Test hemmenden Einfluß durch den Tumor selbst und unterschiedliche blockierende Faktoren zurückgeführt [665, 1068].

Die Aussagen der CMC-Methode zur Definition von Tumorantigenen werden in unterschiedlicher Weise durch natürliche, zytotoxische Zellen (Natural killer cells) eingeschränkt, die unabhängig von einer spezifischen Immunität in unterschiedlicher Menge auftreten und manchmal sogar die CMC übertreffen können [774].

Als relativ spezifisch erwies sich die Lymphozytenstimulation durch autologe Tumorzellen oder aber auch durch allogene Tumorzellen des gleichen Typs [231, 279, 1553]. Jedoch muß die Aussagefähigkeit dieser Technik eingeschränkt werden, da falsch-positive Reaktionen aufgrund von Stimulationen durch tumorunabhängige Differenzierungsantigene entsprechend einer autologen MLC-Reaktion durchaus möglich sind [1068].

Insgesamt wurde mit zunehmender Erfahrung mit den unterschiedlichen Testverfahren offensichtlich, daß der In-vitro-Nachweis von zellulär vermittelter Immunität häufig nicht mit den in vivo bestehenden Hinweisen auf eine Immunreaktion gegen Tumoren korrelierte, und in vitro beträchtlich häufiger Kreuzreaktivitäten auftraten als bei den entsprechenden In-vivo-Testungen. Als In-vivo-Test diente hierbei die Transplantation von Tumorzellen im Gemisch mit sensibilisierten Lymphozyten (Winn assay) [1068]. Für die Kreuzreaktionen wurden viruskodierte Membranantigene [173, 1076] oder embryonale Antigene [1136], die zusätzlich zu den TSTA auftreten können, verantwortlich gemacht.

Beim Menschen ergaben sich ähnliche Ergebnisse wie in experimentellen Systemen. So erwiesen sich in ersten Untersuchungen lymphozytotoxische Reaktionen gegen autochthone Tumoren im Vergleich zu Kontrollzellen als spezifisch. Dieses sprach für die Anwesenheit von Tumorantigenen [578, 579, 1044]. Hier konnten sogar Kreuzreaktionen gegen allogene Tumorzellen innerhalb eines gleichen histologischen Typs gefunden werden (Übersicht bei [1068]). Jedoch waren in nachfolgenden Untersuchungen und mit optimierten Nachweistechniken diese Ergebnisse zum Teil nicht zu reproduzieren [603, 1068], oder die Lymphozyten von Kontrollpersonen wie Normalpersonen oder von Patienten mit einer unterschiedlichen Tumorerkrankung erwiesen sich gegen die jeweilige Tumorzelle gleich oder stärker zytotoxisch als die autologen Lymphozyten des Probanden [598, 619, 1079, 1129, 1472].

Für diese widersprüchlichen Ergebnisse können mannigfaltige Ursachen angenommen werden, wie beispielsweise eine Variabilität in der zytolytischen Sensibilität der Zielzellen (frische isolierte Tumorzellen oder Tumorzellkulturen), Unterschiede in der Therapie der Lymphozytenspender, begrenzte Anzahl von zur Verfügung stehenden Lymphozyten, Einfluß der Lymphozytenisolierungsmethode, Modifikationen bei der Durchführung des Zytotoxizitätstests und unterschiedlicher Einfluß von natürlicher Killer-Zell-Aktivität auf die verschiedenen Zielzellen (Übersicht bei [813, 1068]).

Zusammenfassend läßt sich jedoch sagen, daß zellulär immunologische Methoden bislang nur eingeschränkt für die Erfassung von Tumorantigenen anwendbar sind. Ihre Aussagen werden entweder durch mangelnde Reproduzierbarkeit oder durch mangelnde Sensitivität oder Selektivität begrenzt.

Nachweis mit Hilfe von Antikörperreaktionen

Ausgehend von Ehrlichs Vermutung einer Antikörperreaktion gegen Tumoren [337] war bereits seit längerem versucht worden, durch Immunisierung von Kaninchen, Pferd, Ziege, Schaf oder Affe mit Tumorgewebe Antikörper zu erzeugen und mit Hilfe dieser Antikörper tumorspezifische Antigene nachzuweisen (Beispiele hierfür bei [38, 424, 1068, 1452]). Für diesen serologischen Nachweis, zuerst mit heterologen und später auch mit allogenen und autologen Antikörpern, wurden im Laufe der Zeit zahlreiche Methoden entwickelt oder übernommen.

Hierzu zählen:
- die Komplement-abhängige Antikörper-mediierte Zytotoxizität gegen lebende Tumorzellen [501, 584];
- die direkte (ein gegen das vermeintliche Tumorantigen gerichteter Antikörper ist mit Fluoreszenzfarbstoffen, mit Enzymen oder mit einem Isotop markiert) oder die indirekte (ein gegen den Erstantikörper

gerichteter Zweitantikörper ist markiert) Immunfluoreszenz oder Immunenzymmethode, bei welcher der Bindungsort (Zellsuspension; Zellrasen oder Gewebeschnitt) der Antikörper durch das Markierungsmittel sichtbar gemacht wird [38, 147, 554, 972, 992, 1318]. Durch Verwendung von F(ab')$_2$ Fragmenten anstelle intakter Antikörper [1318] konnte die Spezifität dieser Nachweismethode deutlich gesteigert werden;

– die «Mixed hemadsorption (MHA)», bei welcher die Bindung von Antikörpern an Tumorantigene durch einen gegen den Erstantikörper gerichteten Zweitantikörper sichtbar gemacht wird, der gleichzeitig über seine Divalenz auch einen Erythrozyten-Ambozeptorkomplex bindet [372, 968];

– die Komplementbindungsreaktion, bei welcher die komplementabhängige Hämolyse im Indikatorsystem (Schafserythrozyten und Ambozeptor) durch Komplementfixation am Tumorantigen-Antikörper-Komplex verhindert wird;

– der Immunadhärenz-Assay (IA) [1468], bei welchem normale menschliche Erythrozyten (Blutgruppe 0) an Zellmembran-Antigen-Antikörper-Komplement-Komplexe binden und damit spezifische Antigenbindung von Antikörpern bei experimentellen Tumoren sowie bei Humantumoren sichtbar machen [690, 1046];

– der Protein-A-Assay (PA), bei welchem radioaktiv oder Fluoreszeinmarkiertes Protein A von Staphylococcus aureus an das Fc-Teil von IgG-Antikörpern bindet, welche wiederum an Tumorzellen gebunden sind [173, 1030, 1210, 1651];

– die Immunpräzipitationsmethoden, bevorzugt solche in Gelen wie die radiale Immundiffusion [1092], die Immunelektrophorese [504, 1268], die Polyacrylamidgelelektrophorese [385, 1661], der «Antigen-spot»-Test [611] und die «Immuno Western blotting analysis» [715, 1550];

– die Antikörper-abhängige zelluläre Zytotoxizität (ADCC), bei welcher bestimmte Lymphozyten (T-Zellen; K-Zellen; Nullzellen und Makrophagen) durch Bindung über Fc-Rezeptoren an das Fc-Teil von zellmembrangebundenen Antikörpern zytotoxisch werden für die Tumorzellen [719, 1119].

Mit Hilfe dieser Methoden konnten sowohl Antigene auf der Zellmembran von tierexperimentell induzierten als auch natürlich «spontan» oder über eine Virusinfektion entstandenen Tumoren entsprechend ihrem Auftreten verschiedenen Kategorien zugeordnet werden.

Klassifizierung der Tumorantigene

Tumorspezifische Oberflächen («Surface»)-Antigene (TSSA)

Bei chemisch induzierten Tumoren konnten parallel zu dem Nachweis mit Hilfe von Transplantationstechniken (TSTA) individualspezifische Antigene auch durch Antikörper nachgewiesen werden [599, 600], zusätzlich jedoch, und dieses in geringem Ausmaße, gruppenspezifische Antigene, möglicherweise bedingt durch das gleichzeitige Auftreten von embryonalen [56, 963, 1438] oder durch endogene Viren [604] induzierten Membranantigenen. Beide scheinen, zumindest gering, auch als TSTA zu wirken [57, 604].

Eindeutig gruppenspezifische Tumorantigene wurden jedoch bei Virus-induzierten Tumoren gefunden (Übersichten bei [604, 847, 1068, 1583]). Diese als Virus-induzierte «Tumor specific surface antigens» (TSSA) bezeichneten Antigene [1504] waren spezifisch für das jeweilige Tumorvirus und demzufolge bei allen Tumoren nachzuweisen, die durch das jeweilige Virus induziert waren oder mit denen das jeweilige Virus «vergesellschaftet» erschien (Übersicht bei [1583]), so mit Hilfe der Komplementbindungsreaktion erstmals bei Adenovirus-induzierten Tumoren in der Ratte und beim Hamster [669], bei SV40-induzierten Tumoren im Hamster [113], und bei SV40-transformierten Zellen von Kaninchen, Maus, Schwein, Rind und Mensch [112, 530, 1190]. Ähnliches konnte durch Bestimmung zytotoxischer Antikörper auch bei Polyomavirus-induzierten Tumoren gefunden werden [574].

Wie am Beispiel von durch RNA-Tumorviren induzierten Tumoren ermittelt, sind die virusspezifischen TSSA von eventuell zusätzlich auf der Membran auftretenden Virusstrukturproteinen zu unterscheiden [847]. Im Gegensatz zu letzteren scheint zum einen das Auftreten von virusspezifischen TSSA mit der malignen Transformation einer Zelle zu korrelieren; zum anderen gibt es Anhaltspunkte, daß Immunreaktionen gegen virusspezifische TSSA vor dem Auftreten eines Tumors schützen oder eine tumortherapeutische Wirkung haben können [298, 361, 820, 1387].

Derartige Befunde konnten nicht nur in experimentellen Virusmodellen, sondern auch bei natürlich auftretenden RNA-Tumor-Virus-bedingten Erkrankungen wie beispielsweise bei den durch Onkornaviren hervorgerufenen Katzenlymphosarkomen gefunden werden, was als Bestätigung der Immune-surveillance-Theorie gelten dürfte [1400, 1443].

Da beim Menschen die Virusgenese vereinzelter Tumoren wahrscheinlich ist, sollten auch hier virusspezifische TSSA zu erwarten sein; so bei speziellen Leukämien hervorgerufen durch das «Human T-cell leukemia virus (HTLV)» bzw. durch das «Adult T-cell leukemia virus» (ATLV) [568, 634, 989, 1209], welches das ATL-cell associated antigen (ATLA) induziert [634, 1291, 1639], und bei Papillomavirus-induzierten Cervixkarzinomen [1660]. Bei dem mit Epstein-Barr-Virus (EBV)-Infektion vergesellschafteten Burkitts-Lymphom und nasopharyngealen Karzinom konnten virusspezifische Antigene (z. B. das Early antigen (EA)) identifiziert werden, gegen welche vom Patienten Antikörper gebildet werden, deren Titer mit der jeweiligen Tumorprogression korrelierte [590, 591].

Tumorassoziierte Antigene (TAA)

Bei dem größten Teil menschlicher Tumoren, aber auch bei zahlreichen «spontan» entstandenen experimentellen Tumoren steht die Suche nach tumorspezifischen Antigenen erst in ihren Anfängen. Sie wird speziell bei Untersuchungen mit heterologen oder allogenen Antikörpern durch Kreuzreaktionen mit allogenen und xenogenen Antigenen beeinträchtigt. Konsequenterweise wurde die Empfehlung ausgesprochen, zur serologischen Definition von Tumorantigenen bevorzugt autologe Antikörper zu verwenden [1066]. Dieser Empfehlung kann aber nur entsprochen werden, wenn autologes Tumormaterial in genügend großer Menge zur Verfügung steht (beispielsweise über Zellkulturen).

Andererseits bietet heute die Hybridomatechnologie [816] die Möglichkeit, zwar xenogene, jedoch monoklonale und damit monospezifische Antikörper herzustellen, mit deren Hilfe Tumorzellmembranantigene erkannt und definiert werden können. Um ihre eventuelle Tumorspezifität zu beweisen, müssen jedoch alle nur möglichen Spezifitätstestungen gegen Antigene auf Normalgewebe durchgeführt werden. Die Herstellung im xenogenen System mag jedoch den weiteren Vorteil haben, daß auch Tumorantigene erkannt werden, gegen die der autologe Wirt «blind» ist.

Die bislang vorliegenden Untersuchungen zur Spezifitätsprüfung von polyklonalen (konventionellen), autologen, allogenen und xenogenen Antikörpern sowie von xenogenen monoklonalen Antikörpern haben noch keinen eindeutigen Beweis erbringen können, daß TSSA auf «spontan» entstandenen Tumoren existieren [268, 587, 865], da zum einen vermeintliche TSSA auch auf normalen Zellen gefunden werden konnten [30, 140, 509, 599, 662, 898, 970, 1039, 1484, 1605, 1637] und andererseits die Spezifitätsprüfung an normalen Geweben noch nicht bei allen gegen Tumoren gerichteten monoklonalen Antikörpern derart breit durchgeführt wurde, daß eine endgültige Aussage möglich ist. In Anbetracht dieser Tatsache beschränkte man sich bislang auf den Begriff der tumorassoziierten Antigene (TAA) [599], die sowohl bei tierexperimentellen Tumoren (Übersicht bei [62]) als auch auf Humantumoren, vorwiegend mit Hilfe monoklonaler Antikörper aber auch mit autologen Seren nachgewiesen werden konnten, so beispielsweise auf Brustkrebs [236, 1061, 1275, 1276, 1484, 1643], kolorektalen Karzinomen [613, 614, 824, 825, 918, 1044], Lungen (Bronchial)-Karzinomen [141, 142, 154–156, 162, 265, 667, 741, 758, 939, 985, 1013, 1274, 1371], Ovarialkarzinomen [73, 265, 741], Prostatakarzinomen [418, 1038, 1437], Gliomen [1280, 1370], Leiomyosarkomen [292], Melanomen [174, 175, 177, 205, 310, 350, 526, 681, 682, 716, 823, 835, 875, 898, 986, 1605, 1606, 1635, 1637], Neuroblastomen [760, 762, 875], Osteosarkomen [351, 662], Retinoblastomen [875], Teratomen [335], Pankreaskarzinomen [140, 970] und Hodgkin- und Sternberg-Reed-Zellen [1301]. Eine auch nur halbwegs aktuelle und umfassende Übersicht über die speziell mit monoklonalen Antikörpern nachgewiesenen TAA bei den verschiedenen Tumoren zu geben, ist wegen der drastisch wachsenden Zahl an Veröffentlichungen über neue Spezifitäten fast unmöglich.

Am intensivsten untersucht wurde bislang das Melanom. Unter Berücksichtigung aller notwendigen Spezifitätsuntersuchungen und Absorptionen mit Kontrollgeweben konnten wiederholt autologe Antikörper im Tumorträger gegen das eigene Tumormaterial gefunden werden, jedoch schwankte der Prozentsatz positiver Probanden je nach angewandter Testmethode beträchtlich [1068, 1361]. Eine endgültige Aussage über das Vorkommen von TSSA auf Melanomzellen ist somit noch nicht möglich. Mit Hilfe von monoklonalen xenogenen Antikörpern konnten biochemische Strukturen auf Melanomzellen als mögliche TAA von Melanomzellen identifiziert werden, so beispielsweise Glykoproteine vom MW 75 000 und ein Protein vom MW 89 000 [656], Glykoproteine vom MW 94 000 [1606], vom MW 97 000 [174–176, 586], 155 000 [898] und 210 000 [898] oder Glykolipide, speziell Ganglioside [310, 1177, 1637].

In ähnlicher Weise konnten bei anderen Tumoren TAA definiert werden, z. B. beim Bronchialkarzinom [141, 154, 155, 667, 1043], beim

Kolonkarzinom [918], beim Pankreaskarzinom [921], bei der T-Zell-Leukämie [1348] und beim Mammakarzinom [966, 1276].

Unter den TAA haben im zunehmenden Maße Zuckerstrukturen, speziell auch auf Glykolipiden, an Bedeutung gewonnen. Zum Teil stellen diese Glykolipide Ganglioside wie GD_3, GD_2, GM_2 oder GM_1 dar [535, 1043], oder aber es sind Blutgruppenantigene (Forssman-Antigen, Antigene der Blutgruppe A oder Lewis) oder Vorläuferstrukturen von Blutgruppenantigenen wie beispielsweise Fucolipide [535, 1391] oder aber diesen Blutgruppenantigenen ähnliche Strukturen [219, 442, 536]. Modifikationen von Blutgruppenantigenen in Tumorzellen können durch Blockade in der Biosynthese und damit vermehrter Ausbildung von Derivaten [442] oder von Vorstufen unter Verlust der Blutgruppen-ABH-Determinanten entstehen. Eine derartig modifizierte, tumorassoziierte Neosynthese von Blutgruppenantigenen kann beispielsweise zur Ausbildung von Blutgruppe-A-Antigenen auf Tumorzellen in Individuen mit der Blutgruppe 0 oder B oder von Forssman-Antigenen in Forssman-Antigen-negativem Gewebe führen [535]. Insgesamt gesehen waren alle bislang entdeckten tumorassoziierten Antigene auch in unterschiedlichen Normalgeweben, und hier manchmal in geringeren Konzentrationen, nachweisbar. Eine Tumorspezifität konnte bislang nur individual-spezifisch für einzelne seltene Tumorträger nachgewiesen werden, so beispielsweise bei Patienten der Blutgruppe 0 mit einer tumorassoziierten Blutgruppe-A-Struktur [535] oder aber bei B-Lymphozytentumoren mit einer bestimmten, für diesen jeweiligen Tumor charakteristischen Idiotypstruktur der membranständigen Immunglobuline [199, 873]. Eine gewisse Tumor-«Spezifität» kann sich ergeben durch quantitative Unterschiede in der Bildung und/oder Zellmembranexponierung von Antigenen zwischen Tumorzellen und Normalzellen. Zusätzlich können TAA auf Tumorzellen Epitope besitzen, welche sich in ihrer Zusammensetzung oder Konformation von ähnlichen auf Normalzellen desselben Gewebes unterscheiden.

Derartig tumorspezifische Epitope mögen das Ergebnis einer inkompletten Synthese oder «posttranslational modification» von antigenen Determinanten der normalen Zelle sein [101, 320]. Naturgemäß sind derartige tumorspezifischen Epitope nur sehr schwer aufzufinden. Sie könnten jedoch die Möglichkeit zumindest einer spezifischen Diagnostik von Tumoren bieten. Vorläufige Ergebnisse von Untersuchungen unserer Arbeitsgruppe weisen in der Tat darauf hin, daß beispielsweise Epitope auf dem CEA gastrointestinaler Tumoren auftreten können, welche in einem hohen Grad mit Tumoren assoziiert sind und nicht auf dem NCA 55 oder NCA 95 vorkommen [144].

Embryonale Antigene

Embryonale oder fetale Antigene werden definitionsgemäß während der Onkogenese synthetisiert oder exprimiert und, falls überhaupt, nur in geringen Mengen beim gesunden erwachsenen Individuum [9, 51, 1505–1507]. Gold et al. [480] und Krupey et al. [838] konnten als erste embryonale Antigene in Tumoren nachweisen, speziell das karzinoembryonale Antigen (CEA), von welchem sie annahmen [479], daß es in Tumoren durch eine «depressive Dedifferenzierung» gebildet wird. In der Folgezeit wurden weitere onkofetale Antigene (α-Fetoprotein; SP_1; PAIP und andere) nachgewiesen (Übersicht bei [56, 220, 789, 852, 853]), jedoch nicht nur bei Tumoren, sondern auch bei anderen Erkrankungen.

Embryonale Zellmembranantigene konnten unabhängig von der Art der Tumorinduktion [56, 1505] und auch bei Virus-induzierten Tumoren mit serologischen Methoden und mit Hilfe von Zytotoxizitätsbestimmungen gefunden werden [187, 234, 325, 1115, 1505–1507]. Immunisierung mit embryonalen Antigenen erbrachte entweder Schutz gegen den Tumor [57, 234], zeigte keinen Einfluß [1505–1507] oder bewirkte sogar eine Beschleunigung des Tumorwachstums [1115].

Ihre Immunogenität scheint jedoch, wenn überhaupt, insgesamt wesentlich geringer zu sein als diejenige von TSTA [220, 599]. Es gibt Anhaltspunkte für die Existenz von organspezifischen embryonalen Antigenen, welche mit Tumoren, die in diesen Organen entstehen, kreuzreagieren [55, 599, 1598].

Organ- bzw. gewebespezifische Antigene

Zellen oder Gewebe können entsprechend ihrer Differenzierung [1069] zytoplasmatische oder membranständige Komponenten und/ oder Sekrete produzieren, die für das jeweilige Organ charakteristisch sind und möglicherweise in der Zellorganisation, der Wachstumskontrolle und bei der Funktion des Gewebes eine Rolle spielen [184, 186]. Ein Beispiel für ein organspezifisches Antigen ist das Thyreoglobulin

[319]. Tumoren eines Organs können manchmal an organspezifischen Antigenen erkannt werden, z. B. Plasmazelltumoren an der Immunglobulinsekretion [509] und die saure Prostataphosphatase beim Prostatakarzinom [877].

Besondere Bedeutung haben Membran-Differenzierungsantigene der verschiedenen Lymphozytenuntergruppen [1192, 1501] bei der Klassifizierung der unterschiedlichen, vom lymphatischen Gewebe entstammenden Tumoren gewonnen [104, 150, 270, 567, 768, 808, 829, 891, 1020, 1123, 1492, 1501]. Versuche, mit Hilfe von monoklonalen Antikörpern gegen Differenzierungsantigene auf Naevi eine Korrelation zwischen dem Stadium und der Malignität von Melanomen einerseits und der Exprimierung von Differenzierungsantigenen andererseits aufzustellen, schlugen jedoch fehl [1462].

Des weiteren wurde versucht, mit Hilfe von Antikörpern gegen Keratin epitheliale Tumoren, gegen Vimentin mesenchymale Tumoren, gegen Desmin Rhabdomyosarkome und gegen Neurafilamente Neuroblastome zu erkennen und voneinander zu unterscheiden [417, 860, 996, 1525]. Jedoch wurden hier deutliche Kreuzreaktionen zwischen den einzelnen Tumorgruppen zum einen und den Filamenten zum anderen gefunden [503, 1173].

Kryptische Antigene

Kryptische Antigene werden definitionsgemäß bei Normalzellen durch terminale oder assoziierte Strukturen maskiert. Beispiele hierfür sind die Maskierung von Antigendeterminanten des β_2-Mikroglobulin durch Assoziation mit HLA-Strukturen [874], die Abdeckung des sogenannten Thomsen-Friedenreich-Antigens (β-D-Galactosyl(1−3)-αN-Acetyl-D-Galactosamin-Peptid) ([422, 423, 1496], Übersicht [1182, 1533]), durch endständige N-Acetyl-Neuraminsäure und die Maskierung des Forssman-Antigens (α-GalNAc-(1−3)-β-GalNAc-(1−3)-α-Gal-(1−4)β-Gal(1−4)-β-Glc-(1-1)-Ceramid) durch endständige N-Acetyl-Neuraminsäure, GalNAc-(1−3)-β-Gal oder Fukose [1057]. Kryptische Antigene werden entweder durch enzymatische Abspaltung der maskierenden Gruppen frei, oder aber die maskierende endständige Gruppe fehlt aufgrund einer unvollständigen Synthese von Glykokonjugaten, wie sie beispielsweise in Tumorzellen, möglicherweise auch als Ergebnis einer Onkogen-Aktivierung vorkommt [534]. So konnte das Thomsen-

Friedenreich-Antigen serologisch und mit Erdnußlektin beim Karzinom der Lunge, des Pankreas, der Harnblase, beim Melanom [1428–1430] und der Mamma [25, 664, 802, 1037, 1423, 1431, 1491] und mit Erdnußlektin beim Kolonkarzinom [245, 1065] nachgewiesen werden sowie das Forssman-Antigen bei menschlichen lymphatischen Leukämiezellen [1418] und Teratokarzinomzellen der Maus [1446], jedoch auch beim Influenza-Virus [1418]. Es bestehen Anhaltspunkte, daß bestimmte Tumorträger gegen das Thomsen-Friedenreich-Antigen eine verstärkte Immunreaktion entwickeln, welche mit serologischen Methoden und mit Hilfe der DTH-Reaktion erfaßt werden kann [432, 1429, 1491]. Kontrolluntersuchungen lassen vermuten, daß diese Immunreaktion durch das von den jeweiligen Tumorzellen unmaskiert präsentierte Thomsen-Friedenreich-Antigen verursacht wird.

Tumorantigene und Histokompatibilitätsantigene

Es werden genetische, strukturelle und funktionelle Beziehungen zwischen TAA und Histokompatibilitätsantigenen (HA) diskutiert [27, 389, 925, 1107]. Für eine genetische Beziehung spricht, daß bei Virus- oder chemisch induzierten Tumoren das Auftreten von TSTA mit dem Verschwinden von HA einherging [569, 1504], des weiteren, daß die Isolierung von Tumorzellen mit gleichem HA-Muster zur gleichzeitigen Gruppierung von bestimmten TAA führte [1090].

Mit Hilfe von zellulären Immunreaktionen gegen Melanome und Mammakarzinome konnten sogar TAA identifiziert werden, welche strukturelle Ähnlichkeit mit HLA-A- und HLA-B-Antigenen besaßen (Polypeptide mit einem Molekulargewicht von 45 000, assoziiert mit β_2-Mikroglobulin [994, 1497, 1498]).

Diese Befunde blieben nicht unwidersprochen [149, 203, 214, 263, 798]. So konnten beispielsweise auch mit monoklonalen Antikörpern bislang keine strukturellen oder räumlichen Beziehungen zwischen TAA und HA bei verschiedenen Tumoren ermittelt werden [681, 1039]. Im Falle des Melanoms scheinen TAA und HLA-A, B-Antigene unterschiedlichen Regulationsmechanismen unterworfen zu sein, da Interferonbehandlung die Expression von HLA-Antigenen erhöht, ohne diejenige von TAA zu beeinflussen [683]. Neuere Befunde über eine Homologie zwischen dem HTLV-envelope-Gen und dem HLA (MHC-Klasse I)-Gen [226] weisen jedoch durchaus auf die Möglichkeit einer genetischen Beziehung zwischen HLA und TAA bei bestimmten Tumoren hin.

Eine weitere Vermutung in diesem Zusammenhang ist, daß in Normalzellen reprimierte Gene, die allogene Histokompatibilitätsantigene kodieren, durch den Prozeß der malignen Transformation aktiviert werden und zur Synthese von neuen «Produkten» führen, welche TAA darstellen können [687]. Für diese Hypothese gibt es sowohl stützende [923, 1243] als auch widersprechende Ergebnisse [450, 451, 924, 1108, 1109, 1215, 1216].

Zumindest für einige Tumoren ist eine funktionelle Beziehung zwischen TAA und Histokompatibilitätsantigenen erwiesen, und zwar speziell unter dem Aspekt der Zytolyse von verschiedenen Tumorzellen durch spezifisch zytotoxische T-Lymphozyten (CTL). Diese Immunantwort wird durch die Präsentation der TAA in Verbindung mit MHC-Gen-Produkten reguliert ([798, 1600]; Übersicht bei [315]). Jedoch gibt es für diese sogenannte MHC-Restriktion neben bestätigenden [239, 487, 1518] auch widersprechende Befunde [644, 1508], speziell solche, die die Regulation der Immunantwort gegen TAA bei manchen Tumoren außerhalb des MHC finden [1620, 1621].

Tumor und Immunsystem

Unser heutiges Wissen über die Wechselbeziehung zwischen Tumor und Immunsystem basiert auf folgenden Erkenntnissen: Lymphozyten sind in B- und T-Zellen zu unterteilen [490, 978]; zellulär-immunologische Vorgänge werden von T-Zellen vermittelt [1577]; die Immunreaktionen werden von T-Helfer-Zellen [745, 988, 1185], von «Suppressor»-T-Zellen [459, 738] und von Makrophagen [1541] reguliert; die gesamte Immunantwort wird von «Immune response» (IR)-Genen, lokalisiert im «Major histocompatibility complex» (MHC-Klasse II), gesteuert [86, 488, 868, 946]. Diese sind gekoppelt an diejenigen (MHC-Klasse-I-) Gene, die die Transplantationsantigene kodieren [551, 947]. MHC-Klasse-II-Gene kontrollieren hierbei nicht die Erkennung des einzelnen Epitops (Hapten), sondern des die Epitope tragenden Trägers (Carriers) [869]. Dieses gilt sowohl für die zelluläre Immunantwort als auch für die Antikörperantwort, soweit sie der Hilfe durch T-Zellen bedarf [86, 326]. Eine T-Zell-unabhängige direkte Aktivierung von B-Lymphozyten durch bestimmte Antigene und die hierdurch resultierende Antikörperantwort unterliegt demnach nicht der Kontrolle der MHC-Klasse-II-Gene [86]. MHC-Klasse-II-Gene beeinflussen T-Zell-abhängige Immunreaktionen über Makrophagen bei der Selektion und Präsentation von Antigenen, über T-Zellen bei der Ausbildung von T-Zellrezeptoren und der Makrophagen-T-Zell-Interaktion und über B-Zellen im Rahmen der T-Zell-B-Zell-Kooperation [85, 489]. Endergebnis dieser Immunreaktion ist die Ausbildung von spezifischen Effektorzellen und spezifischen Antikörpern. Neben dieser antigenspezifischen Immunreaktion gegen Tumorzellen bestehen in Form von zytotoxischen Makrophagen, natürlichen Killer (NK)-Zellen und Granulozyten nicht-antigen-spezifisch ablaufende Abwehrmechanismen.

Experimentelle Daten über den Anstieg der Metastasierung im Gefolge einer allgemeinen Immunsuppression [204, 405, 920, 1351] oder einer spezifischen Suppression von T-Zellen [271, 331] und Makrophagen [210, 723] oder bei einer Defiziens von natürlicher Killer-Zellaktivität [1477] weisen auf die mögliche Rolle der unterschiedlichen Komponenten des Immunsystems bezüglich des Wachstums von Tumoren und ihrer Metastasen hin.

Antigenspezifische Immunabwehr

Erkennung von Antigenen, inklusiv von Tumorantigenen (afferent)

Seit Anfang der siebziger Jahre ist bekannt, daß Lymphozyten nur dann eine Immunantwort gegen ein Antigen entwickeln können, wenn dieses Antigen zuvor von Makrophagen in einer für Lymphozyten erkennbaren Form zubereitet worden ist [1537]. Neben Makrophagen gehören zu diesen «Antigen presenting cells» (APC) die Langerhans-Zellen in Haut, Oesophagus und Cervix [153, 1372, 1450], die dendritischen Zellen in Milz, Lymphknoten, Peyersche Platten und Blut [164, 1176], die lymphofollikulären dendritischen Zellen [788], die «Interdigitating» Zellen in parakortikalen Bezirken der Lymphknoten und in T-Zell abhängigen Gebieten der Milz und im Thymus [62], Endothelzellen [637] und Monozyten [489]. Die APC konnten in der Maus, Ratte, Kaninchen, Meerschweinchen und Mensch gefunden werden [1147] und scheinen ihren gemeinsamen Ursprung im Knochenmark zu haben [673]. Die spezifische Funktion der APC wurde größtenteils in vitro mit markierten Antigenen ermittelt, und von diesen Ergebnissen wurde auf die In-vivo-Situation rückgeschlossen [1147]. In-vivo-Experimente nach Injektion unterschiedlicher APC und gleicher Antigene ergaben, daß Unterschiede in der «Antigen presenting»-Aktivität unterschiedlicher APC bestehen [164, 1176]. Untersuchungen an Nierentransplantaten in der Ratte [862] und am Menschen [1180] weisen auf eine wesentliche Rolle der dendritischen Zellen bei Einleitung der Organabstoßung hin. Da Parallelen zwischen der Organabstoßung und der Tumorabwehr zu bestehen scheinen, könnten auch hier dendritische Zellen initiale Funktionen darin besitzen, TSTA, TSSA oder TAA derart zu verarbeiten und zu präsentieren, daß T-Lymphozyten stimuliert werden [1540]. Ähnliche

Funktionen könnten alle anderen APC in unterschiedlichem Ausmaß erfüllen. Für die Stimulation von T-Lymphozyten ist dabei die Anwesenheit von Immune-response-MHC-Klasse-II-Gen-kodierten Antigenen (= Ia Antigene der Maus und HLA-DR Antigene des Menschen [250, 489, 947, 1247]) auf der Zellmembran von APC notwendig [454, 1127, 1305]. Da lymphofollikuläre dendritische Zellen diese Antigene nicht tragen, sind sie möglicherweise für die T-Lymphozytenstimulation weniger geeignet. Es bestehen Anhaltspunkte, daß lymphofollikuläre dendritische Zellen Antigene über Monate und Jahre hinweg auf ihrer Oberfläche halten und hierdurch an der Regulierung der Antikörperantwort teilhaben [1149]. Im Gegensatz zur Aktivierung von T-Lymphozyten scheint die Stimulation von B-Lymphozyten durch T-Zell-unabhängige Antigene nicht die Anwesenheit von MHC-Klasse-II-Gen-kodierten Antigenen auf APC zu benötigen [673]. APC, wie dendritische Zellen und Makrophagen, potenzieren die Primär- und Sekundärreaktion von T-Lymphozyten auf zelluläre [926, 1059, 1441, 1442] und synthetische Antigene [1060, 1461]. Es ist unklar, in welcher detaillierten Weise APC das Antigen erkennen, verarbeiten und präsentieren [2]. Vermutlich sind die MHC-Klasse-II-Gen-kodierten Antigene auf der Zellmembran von APC in der Lage, sich mit bestimmten Aminosäurensequenzen von Antigenen zu verbinden und dadurch einen membranständigen Komplex zu bilden, der wiederum von bestimmten Rezeptoren als ein autologes MHC-Klasse-II-Gen-kodiertes Antigen und als ein Antigen auf der Zellmembran von speziell differenzierten T-Lymphozyten erkannt wird [84, 1222]. Eine andere Hypothese nimmt an, daß unter genetischer Steuerung durch MHC-Klasse-II-Gene T-Lymphozyten im Rahmen der Bildung von Rezeptoren für autologe MHC-Antigene auch solche für Antigene ausbilden [1562]. Verstärkte Phagozytose scheint für die Antigenpräsentation nicht ausschlaggebend zu sein, da sie dendritischen Zellen fehlt [1558]. Bei Makrophagen ist bekannt, daß zwar nicht die Bindung, aber die Aufnahme zur Präsentation eines Antigens ein energieverbrauchender Prozeß ist [348, 1542, 1570], der mit einer verstärkten Bildung von MHC-Klasse-II-Gen-kodierten Antigenen auf der Zellmembran dieser Makrophagen und mit einer Sekretion des T-Zellen-aktivierenden Faktors (Interleukin I) einhergeht. Die Expression von MHC-Klasse-II-Gen-kodierten Antigenen auf Makrophagen ist demnach ein variabler Prozeß; negative Makrophagen werden durch Einfluß von Lymphokinen oder nach der Phagozytose oder unter Kulturbedingungen positiv [1031]. Im Gegensatz hierzu tragen dendritische Zellen

MHC-Klasse-II-Gen-kodierte Antigene in stabiler Form [1440]. Zur Aktivierung von T-Lymphozyten scheint die direkte physikalische Bindung von dendritischen Zellen oder von Makrophagen an T-Lymphozyten notwendig zu sein [1226, 1442]. Antikörper gegen das Antigen inhibieren diesen Prozeß nicht [348, 1542].

Im Gegensatz hierzu hemmen Antikörper gegen MHC-Klasse-II-Gen-kodierte Antigene sowohl die Aktivierung von T-Lymphozyten [93, 160, 352] als auch die Adhäsion zwischen Makrophagen und T-Zellen [965, 1358, 1360]. Nicht nur die Expression der für die Stimulierung notwendigen MHC-Klasse-II-Gen-kodierten Antigene auf Makrophagen, sondern auch ihre entsprechenden komplementären Rezeptorstrukturen auf T-Lymphozyten [1127, 1225] werden durch MHC-Klasse-II-Gene kontrolliert, zusätzlich jedoch auch die «Leistung» des Makrophagen, bestimmte Epitope eines Antigens für die Erkennung durch T-Zellen auszuwählen, was letztendlich die Induktion einer T-Zell-mediierten Immunantwort gegen verschiedene Antigene einer genetischen Kontrolle unterwirft [67, 488, 1222, 1223, 1359, 1539]. Die Antigenpräsentation durch APC reicht jedoch nicht für die T-Zell-Stimulierung aus. Zusätzlich ist der Einfluß eines Lymphozyten(T-Zell)-aktivierenden Faktors (Interleukin I) notwendig, den sowohl dendritische Zellen [1442] als auch aktivierte Makrophagen [990] sekretieren. Makrophagen werden wiederum zu dieser Sekretion durch T-Lymphozyten stimuliert, entweder bei dem Zell-zu-Zell-Kontakt im Verlauf der MHC-Klasse-II-Gen-restringierten Antigenpräsentation und damit unter der Voraussetzung einer Ia- bzw. HLA-DR-Antigenidentität [1539] oder aber durch Lymphokine, beispielsweise den «Colony stimulating factor» (CSF) [961, 998].

Die antigenspezifische Stimulierung durch APC und Interleukin I stimuliert T-Helfer-Zellen zur Sekretion des «T-cell growth factors» (Interleukin II), unter dessen Einfluß spezifisch geprägte T-Vorläuferzellen zu T-Effektorzellen heranreifen [486, 1395, 1406, 1456]. Die antigenspezifische Prägung der T-Effektorzellen kann durch antigenspezifische Helferfaktoren erfolgen [572, 1469] oder aber durch einen direkten Kontakt zwischen Helferzelle und Effektorzelle, vermittelt entweder über das Antigen [1185] oder aber über das Idiotypen-Antiidiotypen-Erkennungssystem [341, 343, 704]. Der Aktivierung von Effektorzellen wirken Suppressorzellen entgegen, aktiviert über aktivierte Helfer-Zellen [196] oder möglicherweise über suppressogene Determinanten des Antigens [1024]. Suppressor-Zellen wirken über antigenspezifische

[1547] oder nicht-antigenspezifische Faktoren [572] oder aber durch direkten Zellkontakt, vermittelt über das Antigen [183, 1349] oder über das Idiotypen-Antiidiotypen-Erkennungssystem [342, 343].

Letztendlich resultieren als Effektorzellen zytotoxische T-Lymphozyten (CTL) und T-Lymphozyten, welche die Immunreaktion vom verzögerten Typ (T-DTH = Delayed type hypersensitivity) mediieren. Des weiteren sind zur spezifischen Immunantwort die Antikörper hinzuzurechnen, die nach Bindung an das Antigen über ihr Fc-Teil Komplement oder aber Killer-Lymphozyten oder zytotoxische Makrophagen zur Antikörper-mediierten Komplement-abhängigen Zytotoxizität («Antibody-dependent cellular cytotoxicity» = ADCC) aktivieren.

Alle diese Effektorfunktionen werden in ihrem afferenten Schenkel durch MHC-Klasse-II-Gen-kodierte Antigene reguliert, so neben der Aktivierung von T-Helfer-Zellen [357–359, 695, 1378, 1379] auch die der T-Suppressorzellen [6, 1302], die spezifische Prägung von T-Zellen für die verzögerte Immunreaktion (T_{DTH}) [510, 979–982]; die spezifische Prägung von zytotoxischen T-Zellen [90, 638, 1406, 1500], die T-Zell-B-Zell-Kooperation [746, 1417, 1540] und damit die spezifische (T-Zell-abhängige) Antikörperbildung mit der komplementverbrauchenden und zellulären (ADCC) Zytotoxizität. In allgemeiner Form ausgedrückt unterliegen demnach die direkten Interaktionen zwischen aktivierten T-Lymphozyten oder zwischen aktivierten T-Lymphozyten und anderen Immunzellen der MHC-Klasse-II-Gen-Kontrolle [87, 489]. Dieses gilt auch für die Sonderformen der T-Zell-vermittelten Immunreaktion wie der Kontaktallergie [1450] (wobei hier die Langerhans-Zelle die wesentliche APC darstellt) und für die Sekretion von Lymphokinen im Verlauf der Lymphozyten-Interaktionen [1540]. Einschränkend ist jedoch zu sagen, daß diese MHC-Klasse-II-Gen-Kontrolle durch Restriktion durch allogene MHC-Moleküle ersetzt wird, falls allogene APC das Antigen präsentieren [1021], und daß des weiteren möglicherweise T-Zellen die Rolle von APC im Falle der CTL-Entwicklung gegen allogene Zellen spielen können und damit auch die Restriktion gegen allogene MHC-Determinanten festlegen [1007].

Auf welchem afferenten Wege Immunreaktionen gegen Tumoren entstehen, ist im einzelnen noch unklar. Im Rahmen der MHC-Klasse-II-Gen-Restriktion müßten T-Zellen, die mit einem Autoantigen (gekoppelt an einem MHC-Klasse-II-Gen-kodierten Antigen) reagieren, im Thymus eliminiert werden, um Autoreaktionen in der Peripherie zu vermeiden [314, 315, 488, 897, 1657]. In der Peripherie könnten Autoreaktionen jedoch gegen fremdgewordenes «modifiziertes» Autoantigen entstehen. Da Autoimmunreaktionen gegen Tumoren nachweisbar sind, spricht dieser Nachweis für die Existenz von Tumorantigenen als modifiziertes Selbst. Andererseits ist auch möglich, daß die MHC-Klasse-II-Gene über ihren Einfluß auf T-Helfer und T-Suppressoraktivität eine Toleranz gegen Autoantigene erzeugen, die durch modifiziertes Autoantigen durchbrochen werden kann [489].

Offensichtlich werden zytotoxische Reaktionen gegen allogene Zellen durch HLA-DR-Antigene beeinflußt [1650], und auch die Abstoßung von transplantierten Tumoren unterliegt der MHC-Klasse-II-Gen-Kontrolle [1109].

In welchem Maße Makrophagen durch Antigenerkennung und -präsentation an der Induktion einer spezifischen Immunantwort gegen Tumoren teilhaben, ist ungeklärt. Eine Stimulierung von T-Zellen durch direkten Kontakt mit Tumorzellen ist nicht auszuschließen [753, 1266], jedoch scheint die Anwesenheit besonders von syngenen, weniger gut von allogenen Makrophagen deutlich stärker T-Zellen zu aktivieren, und diese Aktivierung scheint sowohl in vitro als auch in vivo an der Proliferationshemmung von Tumoren meßbar zu sein [1514, 1517]. Hierbei wird die Funktionssteigerung der T-Lymphozyten entweder auf die Antigenpräsentation durch Makrophagen [1125, 1631] oder aber auf Antigen-unabhängige stimulierende Einflüsse der Makrophagen (beispielsweise Sekretion von Interleukin I) zurückgeführt [1620, 1621].

Zelluläre Abwehrreaktionen (zytotoxische T-Zellen und Reaktionen vom verzögerten Typ)

Tumoren sind häufig von lymphoiden Zellen infiltriert und dieses sowohl bei experimentellen Tumoren [289, 449, 469, 763, 1133, 1241, 1654] als auch bei Humantumoren [307, 349, 688]. Allgemein werden diese Lymphozyteninfiltrationen unter dem Blickwinkel einer zytotoxischen Reaktion gegen autologes Tumormaterial als prognostisch günstiges Zeichen für die Tumorerkrankungen angesehen [1556]. In den Zellinfiltraten können vorwiegend T-Zellen [445, 562, 721], auch in einem aktivierten Stadium [445] und hier speziell CTL [722], nachgewiesen werden, jedoch auch NK-Zellen, ADCC-Zellen und Makrophagen (Übersicht [94]). Den CTL wird eine wesentliche Rolle bei der antigenspezifischen Immunreaktion gegen Tumoren zugeschrieben. In Analogie zur MHC-Klasse-II-Gen-Restriktion der T-Zellantwort im afferenten Schenkel gibt es bei der Erkennung der Tumorzelle durch CTL eine genetische Restriktion, vermittelt durch Transplantationsantigene, das heißt durch Antigenprodukte der MHC-Klasse-I-Region. CTL können nur dann ein Fremdantigen erkennen und die das Fremdantigen tragende Zelle lysieren, wenn auf der Zellmembran der Zielzelle neben den spezifischen Antigenen Transplantationsantigene vorhanden sind, und diese Transplantationsantigene denjenigen auf den CTL gleichen [102, 1357, 1658, 1659].

Diese MHC-Restriktion zytotoxischer T-Zellen gilt für das autologe und syngene System und wurde im Detail an Virus-infizierten Zellen erarbeitet [1658]. Es konnte auch für künstliche Antigene (chemisch modifizierte Zellen) und in verschiedenen Spezies gefunden werden (Übersicht [1659]). Es ist anzunehmen, daß es auch für die Tumorzellen gilt. Die Transplantationsantigenrestriktion kann als Mechanismus der CTL angesehen werden, selbst von nicht-selbst bzw. verändert-selbst zu unterscheiden [85]. Vielleicht muß unter diesem Blickwinkel auch der Sinn der Entwicklung von Transplantationsantigenen gesehen werden. Bekannt ist, daß CTL stärker gegen allogene denn gegen syngene oder xenogene MHC-kodierte Antigene als Zielantigen reagieren, was die Vermutung unterstützen kann, daß nur solche T-Zellen den Thymus verlassen dürfen, welche eine geringe Affinität zu autologen MHC-Antigenen, jedoch eine hohe Affinität zu «veränderten» MHC-Antigenen haben [185, 703, 864]. Diese Affinität zu «veränderten» MHC-kodierten Antigenen impliziert auch eine relativ hohe Kreuzreaktivität syngen (beispielsweise durch Virus-infizierte Zellen) stimulierter CTL gegen allogene «normale» Zielzellen [85, 401, 1563] und wirft demzufolge auch die Frage nach der Spezifität der CTL-Aktivität auf.

Gemeinsam mit der Erkennung des Transplantationsantigens erfolgt auch die Erkennung des Fremdantigens, speziell des Tumorantigens. Hierfür besitzen die T-Zellen Antigenrezeptoren, deren Strukturen der variablen (antigenbindenden) Region der schweren Kette eines Immunglobulins ähnlich sind [243]. Die Gene, welche die schwere Kette des Antikörpers kodieren, sind gekoppelt an diejenigen, welche den T-Zell-Antigenrezeptor kodieren (Übersicht [371]).

Für die Erkennung der Tumorzelle durch CTL ist der Zell-zu-Zell-Kontakt notwendig [1218, 1603], wobei es unwahrscheinlich ist, daß die hierbei auftretende unterschiedliche Bindungsstärke einen Einfluß auf die Kinetik der Zytolyse hat [94]. Der Kontakt zwischen den CTL und der Tumorzelle führt nicht zur Ausbildung von Zytoplasmabrücken [730, 1246, 1255]. Zur Zytolyse ist jedoch ein bestimmter Membranzustand notwendig, da Lokalanästhetika mit ihrem bekannten Einfluß auf Magnesium und Kalziumionen und auf die Lipiddoppelschicht der Zellmembran die Bindung von CTL und die Lyse von Tumorzellen inhibieren können [759]. Einen ähnlichen inhibitorischen Effekt haben Cholesterol und Phospholipide [96].

Für den zytolytischen Vorgang werden unterschiedliche Faktoren verantwortlich gemacht, wie Leckbildung durch den Zell-zu-Zell-Kontakt mit Ausfluß von K-Ionen und Einfluß von Na-Ionen [97], Scherkräfte [1337], Sekretion hydrolytischer Enzyme [1644] und/oder Aktivierung von membranständiger Phospholipase, die eine Fettsäure vom Phosphatidylcholin abspaltet, was zur Bildung des zytolytischen Detergens Lysophosphatidylcholin (Lysolecithin) führt [433]. Verschiedene Inhibitionsexperimente lassen es jedoch als fraglich erscheinen, ob Phospholipasefreisetzung das zytolytische Prinzip von CTL ist [94, 95]. Eine andere Möglichkeit ist die Ausschüttung von Lymphotoxin. Lymphotoxin wird von T-Zellen nach Stimulation mit Antigenen, Mitogenen und Proteasen sekretiert [806, 941, 1239]. Es hat vielfältige Wirkungen. Lymphotoxin wirkt zytotoxisch [369, 1221, 1262, 1599] auf verschiedene Zellen einschließlich Tumorzellen [1219].

Lymphotoxin kann des weiteren im Zuge der NK-Zell-mediierten Zytolyse nachgewiesen werden [1625], jedoch ist unklar, ob es am zytolytischen Prozeß direkt teilnimmt oder nur die Zielzellen für die Zytolyse sensibilisiert [364].

In der Tat bestehen Anhaltspunkte dafür, daß Lymphotoxin die Sensibilität von Tumorzellen für die Zytolyse durch NK-Zellen erhöht [364]. Lymphotoxin inhibiert direkt das Wachstum von syngenen und isogenen Tumoren und das Wachstum von xenogenen Normalzellen [369]. Diese Inhibition ist irreversibel und eingeschränkt Spezies-spezifisch. So wirkt menschliches Immuntoxin auf menschliche und Mäuse-Tumorzellen, nicht jedoch auf Tumorzellen des Meerschweinchens. Andererseits wirkt Immuntoxin vom Meerschweinchen auf Tumorzellen vom Meerschweinchen und der Maus, nicht aber des Menschen [1227].

Für die Wirkung scheint die Bindung des Lymphotoxins an β-Galaktosylgruppen auf der Membran der Zielzelle notwendig zu sein, da Galaktose sowie Phythämagglutinine die Wachstumsinhibition durch Lymphotoxin hemmen können [815, 1263]. Des weiteren scheint die durch Karzinogene induzierte maligne Transformation in vitro [368] und in vivo [1187] durch Lymphotoxine inhibiert zu werden.

Die Inhibition der Karzinogenese verläuft über einen noch nicht im einzelnen bekannten, nicht-zytotoxischen Mechanismus [368], der von einem vorübergehenden Anstieg in der Glukosaminaufnahme und in der Synthese hochmolekularer Membranglykoproteine begleitet wird [434]. Nach Karzinogenexposition kann auch die Promotion (z. B. durch Phorboldiester) durch Lymphotoxin inhibiert werden [308, 309]. Lymphotoxin kann des weiteren die Interferonsekretion [1463] und dadurch indirekt die NK-Zellaktivität stimulieren. Die Produktion von Lymphotoxin unterliegt einer Rückkopplungskontrolle [1238].

In ersten vorläufigen klinischen Untersuchungen bestehen Anzeichen für einen zumindest lokalen tumortherapeutischen Effekt [803, 1551]. Trotz dieser erwiesenen Wirkungen von Lymphotoxin ist unklar, ob die Zytolyse durch CTL wirklich über die Ausschüttung von Lymphotoxinen vermittelt wird. So konnte durch Antikörper gegen Lymphotoxin zwar die zelluläre Reaktion vom verzögerten Typ inhibiert werden, was wiederum Einblick in die DTH gab, jedoch war die zytolytische Reaktion nicht zu inhibieren [452]. Dennoch ist bislang nicht auszuschließen, daß Lymphotoxin im Rahmen des Zell-zu-Zell-Kontaktes durch spezielle Membranverbindungen in die Zielzelle gelangt [94, 95]. Durch Bindung von Antigen an spezifisch geprägte T-Zellen werden zusätzlich zum Lymphotoxin noch weitere Lymphokine freigesetzt, so Immun- oder Interferon- und der Makrophagen-aktivierende Faktor (MAF). Aktivierte Makrophagen sekretieren unter anderem Interleukin I, welches wiederum T-Helfer-Zellen zur Sekretion von Interleukin II, und dieses T-Vorläuferzellen zur Proliferation und Differenzierung in T-Effektorzellen stimuliert.

Zu diesen Effektorzellen gehören auch T-Zellen, verantwortlich für die verzögerte zelluläre Immunantwort (T_{DTH}). Die Aktivierung dieser T_{DTH} wird durch MHC-Klasse-II-Genprodukte kontrolliert, das heißt für die Erstauslösung einer DTH sind Makrophagen und/oder andere APC (in der Haut beispielsweise Langerhans-Zellen) notwendig, welche das Antigen gemeinsam mit dem MHC-Klasse-II-Genprodukt T-Helfer-Zellen präsentieren (Übersicht [1540]) und über diesen Weg schlußendlich auch T_{DTH} stimulieren. Die resultierende Entzündungsreaktion vom verzögerten Typ entsteht aus einem Gemisch von Lymphozyten- und sekundären Makrophagenreaktionen [1148]. Sie kann durch Antikörper gegen Lymphotoxin inhibiert werden [452] oder durch Verarmung des Organismus an Makrophagen [948] oder aber auch durch Faktoren, welche die Makrophagenaktivität hemmen [1033]. Derartige Faktoren können auch von Tumoren sekretiert werden [1033].

Andererseits gibt es Anzeichen dafür, daß eine DTH-Reaktion, in der Nähe eines oder im Tumor selbst unter Verwendung eines nichtverwandten Antigens ausgelöst, diesen zur Einschmelzung bringen kann [1399, 1530].

Humorale Abwehrreaktionen

Für die Antikörper-mediierte Komplement-abhängige Lyse von Tumorzellen ist neben dem spezifischen Antikörper ein qualitativ und quantitativ funktionsfähiges Komplementsystem notwendig. Bei Gabe von Antikörpern, spezifisch für TAA oder TSSA zur passiven Therapie von bestimmten Tumorerkrankungen wie beispielsweise der Leukämie bzw. dem Lymphosarkom der Katze [555] oder der AKR-Leukämie [742] hat sich speziell die Notwendigkeit quantitativ ausreichender Komplementfaktoren ergeben [323]. Mangelnde therapeutische Wirkung der Antikörper konnte durch Substitution mit mangelnden Komplementfaktoren [742] oder mit Frischserum als Komplementquelle behoben werden [555].

Durch Bindung der Antikörper an die Zielzelle kommt es zu einer größeren Dichte an Fc-Teilen, welche Komplement über den alternativen oder (vorwiegend) klassischen Weg aktivieren und letztendlich zur Ausbildung des $C_{5b}-9(m)$ Komplexes führen (Übersichten bei [382, 1010]). Dieser Komplex hat eine Hohlzylinderform und penetriert in die apolare «Bilayer»-Schicht der Zellmembran ein, was zu deren Öffnung und damit zur Zell-Lyse führt (Übersicht bei [105]).

Zum anderen werden durch Zellmembran gebundene Antikörper weitere Plasma-Enzymsysteme wie der Hagemann-Faktor [867] oder über direkte und indirekte Wechselwirkungen mit Fc-bzw. Komplementrezeptoren bzw. über Komplementspaltprodukte Granulozyten und Thrombozyten zur Exozytose stimuliert. Derartig freigesetzte lysosomale Faktoren aktivieren wiederum das Hagemann-System, das Gerinnungssystem und das fibrinolytische System, und führen demzufolge zumindest lokal zu einer Entzündungsreaktion (Übersicht [1313, 1314]), welche synergistisch, aber auch antagonistisch zur zellulären spezifischen Immunreaktion wirken kann. So ist beispielsweise eine hemmende Wirkung von C_3 und von C_3-Spaltprodukten auf die Entstehung von CTL beschrieben worden [64].

Unspezifische Immunabwehr

Die unspezifische Immunreaktion gegen Tumoren kann untergliedert werden in diejenige von Makrophagen [1031], die durch Exozytose und durch ihre Sekretionsprodukte einen direkten Einfluß auf die T-Zell-vermittelte antigenspezifische Immunreaktion ausüben, und in diejenigen Zellsysteme, welche Tumorzellen entweder unmittelbar erkennen und lysieren können (zytotoxische Makrophagen und natürliche Killer-Zellen), oder aber erst durch Vermittlung eines an die Tumorzelle spezifisch bindenden Antikörpers (Antikörper-abhängige zelluläre Zytolyse; ADCC).

Makrophagen

Die wichtigsten Sekretionsprodukte (Übersicht bei [1540]) von Makrophagen lassen sich unterteilen in degradative Enzyme (lysosomale Enzyme) [274], Plasminogen-Aktivatoren [1544], Kollagenase und Elastase [1601, 1602], neutrale Proteasen [496, 564], Esterasen [1597], Fibronektin [15], Komplementfaktoren [91, 242, 850], Lysozyme [497], die Gerinnung beeinflussende Faktoren [1409, 1544], Interleukin I [464], Colony stimulating factor [217, 844, 845], Interferon [1392], Angionesis-Faktor [1138, 1490], Prostaglandine [498, 1310, 1311], Tumor-Nekrosis-Faktor (TNF) [1237] und Sauerstoffradikale [971]. Makrophagen werden zur Exozytose und zur Sekretion durch Aktivierung angeregt [495, 911].

Diese kann phänomenologisch und funktionell in drei Stufen eingeordnet werden: die stimulierten Makrophagen, die aktivierten Makrophagen und die aktivierten zytotoxischen Makrophagen [630]. Aktivierung von Makrophagen geht einher mit Veränderungen im Membranpotential, bedingt durch eine verstärkte Membranpermeabilität für Kaliumionen [447] und mit qualitativen und quantitativen Veränderungen der Zusammensetzung der Membranglykoproteine [1638]. Eine Zunahme gesättigter Fettsäuren in den Membranphospholipiden korreliert mit einer Abnahme in der Endozytoserate [1310]. Makrophagen können mindestens auf drei verschiedene Arten aktiviert werden, die entweder alleine oder in Kombination einwirken: entweder durch indirekte Einwirkung oder aber durch Aktivierung der Komplement-, Gerinnungs- und Plasminogenkaskaden, wobei Wirkstoffe entstehen, welche Makrophagen beeinflussen können, oder aber durch Makrophagen-aktivierende Faktoren nach Stimulation von Makrophagen, von Lymphozyten, Granulozyten oder anderen Zellen, die Makrophagen-aktivierende Faktoren freisetzen können [1178].

So kann eine Aktivierung im Rahmen der antigenspezifischen Interaktion mit T-Helfer-Zellen durch Lymphokine eintreten (Makrophagen-aktivierender Faktor MAF), jedoch auch durch verschiedene Immunstimulanzien und Bakterien induziert werden (Übersicht [1051]). Auch Lymphozyten und Fibroblasteninterferon aktivieren speziesspezifisch in vitro und in vivo Makrophagen zur Phagozytose [317, 666, 786] und zur Zytotoxizität [318, 1293, 1295, 1296]. Die Aktivierung von Makrophagen bewirkt eine unspezifische antimikrobielle Resistenzsteigerung [909, 912, 1377], des weiteren auch eine verstärkte zytotoxische Aktivität gegen transformierte Zellen [625] und Tumorzellen [627]. Beide Phänomene müssen jedoch nicht immer gleichzeitig auftreten [1607].

Interleukin I (LAF) und Prostaglandine sind die wichtigsten Sekretionsprodukte von Makrophagen, welche antigenunspezifisch die T-Zell-mediierte Immunantwort beeinflussen:

Interleukin I [464, 1085] wird sekretiert während der MHC-Klasse-II-Gen-kontrollierten Interaktion mit Thymozyten (Übersicht [1540]), jedoch auch durch den Einfluß von unspezifischen Stimuli wie Endotoxin [120, 464, 848, 961], Latexkügelchen und Immunkomplexen [120] und Immunmodulatoren wie Bestatin [696]. Inhibition der RNA oder Protein-Synthese verstärkt die durch unspezifische Stimuli induzierte Interleukin-I-Sekretion, was auf die Inhibition eines Kontrollproteins hindeutet [1543]. Wie bereits diskutiert (S. 26) stimuliert Interleukin I T-Helfer-Zellen, die ihrerseits Interleukin II produzieren, des weiteren MAF und damit zum einen (Interleukin II) die Differenzierung von T-Vorläuferzellen in T-Effektorzellen einleiten und die T-B-Zellkooperation stimulieren und zum anderen (MAF) weitere Makrophagen aktivieren und damit die Produktion von Interleukin I verstärken.

Makrophagen stellen eine wesentliche Quelle von Zyklo-Oxygenase- und Lipoxygenase-Spaltprodukten (Prostazykline und Leukotriene) der Arachidonsäure dar [1231, 1311]. Prostaglandine (PGE_2 und

PGF$_1$) werden von Makrophagen freigesetzt [275, 457, 498, 843]. Synthese und Freisetzung ist nach Aktivierung der Makrophagen, z. B. durch Zymosan, Antigen-Antikörper-Komplexe oder Aggregate von IgG [126, 275, 672, 1113, 1311], durch Endotoxin [843], MAF und CSF [498, 844] deutlich verstärkt.

Bemerkenswert ist, daß nicht alle Substanzen, die Makrophagen zur Interleukin-I-Sekretion stimulieren, auch zu einer Erhöhung der Prostaglandinsekretion führen. So sind Latexkügelchen beispielsweise ohne Wirkung auf die Prostazyklinsekretion [1113], ebenso eine In-vitro-Behandlung mit Corynebacterium parvum oder BCG [275], während im Gegensatz hierzu bei Makrophagen von Tieren, die mit C. parvum oder BCG behandelt worden waren, entweder die Sekretion von Prostazyklinen deutlich erhöht [1513] oder auch erniedrigt ist [1312]. Prostaglandine hemmen eine Reihe von Aktivitäten innerhalb der Immunreaktion, so beispielsweise die Sekretion von Lymphokinen [498], die Zytotoxizität von zytolytischen T-Zellen [594], die Antikörper-abhängige zelluläre Zytotoxizität (ADCC) [324] und die Entstehung und Aktivität von natürlichen Killer-Zellen [311, 312, 324, 1512, 1513]. Die Hemmung der zytolytischen Aktivität scheint auf den durch Prostaglandine induzierten Anstieg von zyklischen AMP zurückzuführen zu sein [595]. Die immunsuppressive Wirksamkeit von Prostaglandinen kann durch Indomethazin inhibiert werden [971].

Da Makrophagen in scheinbarer Abhängigkeit von ihrem Aktivierungszustand sowohl Lymphozyten-aktivierende (Interleukin I) als auch immunsuppressive Substanzen (Prostaglandine) sekretieren und beide Substanzklassen wiederum entweder fördernd oder hemmend auf die Sekretion des Makrophagen-aktivierenden Faktors (MAF) durch T-Lymphozyten einwirken, besteht somit ein Regelkreis zur nichtantigenabhängigen Kontrolle der antigenspezifischen Immunantwort [498].

Ein weiterer Mediator, welcher außer von Lymphozyten und von Fibroblasten auch von Makrophagen, besonders nach Aktivierung durch Lymphokine, produziert wird, ist Interferon [1392, 1409]. Von den bislang entdeckten unterschiedlichen Interferonen ist bekannt, daß sie neben ihrer antiviralen Wirksamkeit [693] die Teilungsrate sowohl von malignen (Übersicht bei [194, 241, 267, 518, 645, 861, 1511, 1531]) als auch normalen Zellen [884, 885, 954] durch Verlängerung der G1- und S+G2-Phase [63] reduzieren, möglicherweise durch Inhibition der Ornithincarboxylase und Beeinträchtigung der Polyaminsynthese [1432], jedoch ist der genaue Wirksamkeitsmechanismus noch unbekannt

(Übersicht [1511]). Andererseits aktiviert Interferon Makrophagen zur Phagozytose [545, 666, 685] und zur Zytotoxizität [705, 1294]. Es hat des weiteren je nach Testsystem, Dosis und Applikationszeitpunkt in Abhängigkeit zur Antigengabe entweder einen hemmenden oder einen fördernden Einfluß auf die Immunantwort (Übersicht bei [1511]). So werden bei zeitgleicher Gabe die «Plaque forming cells» [713, 714] und die «Mixed lymphocyte reaction» [885] reduziert; nach hoher Dosis wird die Transplantatüberlebenszeit erhöht [290, 686], während geringe Dosen zu einer Beschleunigung der Transplantatabstoßung führen [686]. Die DTH-Reaktion gegen ein T-Zell-abhängiges Antigen (SRBC) wird bei zeitgleicher Gabe von Interferon [291] verstärkt, was durch Inhibition von T-Suppressorzellen bedingt zu sein scheint [812]. Andererseits wirken T-Suppressorzellen bei Anwesenheit von Interferon deutlich stärker suppressiv [618], während die Entwicklung von allospezifischen Suppressor-T-Zellen durch Interferon inhibiert wird [416]. Interferon verstärkt die Zytotoxizität von T-Zellen gegen allogene Antigene [35, 617, 880], hemmt jedoch die zytotoxische Reaktion gegen autologe Tumorzellen [1554]. Interferon verstärkt des weiteren die In-vitro-Aktivität von NK-Zellen [608, 1521], wahrscheinlich durch Einfluß auf den Differenzierungsprozeß [1373]. Jedoch kann diese Interferon-bedingte Aktivitätszunahme von NK-Zellen nur gegen allogene, nicht aber gegen autologe «frische» Tumorzellen beobachtet werden [1555]. Auch die Aktivität der ADCC von Killer-Zellen wird durch Interferon stimuliert [1089] und dieses auch bei Tumorpatienten [134].

Sauerstoffradikale, gebildet von Makrophagen, aber auch von neutrophilen Granulozyten, scheinen zumindest in vitro die Lymphozytenfunktion zu beeinträchtigen. Zu diesen chemisch aktiven Substanzen gehören Superoxydanion (O_2^-); Wasserstoffsuperoxid (H_2O_2) und Singlet Sauerstoff ($^1O_2$) [717, 790, 971]. Diese Sauerstoffradikale bewirken zum einen eine direkte Zellschädigung und scheinen hierdurch beispielsweise bei der Zytolyse von Tumorzellen durch Makrophagen eine Rolle zu spielen [1027, 1028]. Zum anderen gibt es Anhaltspunkte, daß Wasserstoffperoxid einen löslichen immunsuppressiven Faktor aktiviert, der von T-Zellen nach Lectinstimulation (ConA) freigesetzt wird und der die Zellproliferation inhibiert [1128, 1470].

Makrophagen können durch Cokultivierung mit syngenen Tumorzellen oder aber durch Kontakt mit verschiedenen Immunstimulanzien und ohne meßbaren Einfluß der T- oder B-Zellen derart aktiviert werden, daß sie für syngene [627, 628, 1081], allogene und xenogene Tumorzellen [69, 221, 227, 256, 394, 627, 628, 736, 1082, 1287] sowie für transformierte Zellen [625] zytotoxisch werden. Mit dieser Aktivierung gehen eine vermehrte Expression von Fc-Rezeptoren [893, 1200] und eine erhöhte Stoffwechselleistung einher (gemessen an der Nitroblau-Reduktion [570] und am Glukosemetabolismus [782]). Die zytotoxische Aktivität von Makrophagen ist immunologisch nicht spezifisch, jedoch selektiv für Tumorzellen [630]. Die Fähigkeit, Tumorzellen zu zytolysieren, ist nicht in allen Fällen mit einer Aktivierung von Makrophagen gekoppelt: so können auch ruhende Makrophagen Tumorzellen lysieren [2, 960]. Die Entwicklung hin zu zytotoxischen Makrophagen scheint ein Differenzierungsprozeß zu sein, der mit einer Veränderung in der Zusammensetzung und/oder Struktur der Membranlipide einhergeht [212, 213]. Je weiter diese Differenzierung fortgeschritten ist, um so geringer ist der Schwellenwert für eine Aktivierung durch äußere Einflüsse [629, 631, 1584].

In welcher Weise Makrophagen transformierte Zellen oder Tumorzellen von Normalzellen unterscheiden können, ist unklar [630, 1031]. Immunisierungen des Makrophagenspenders mit dem Tumor oder eine starke Immunogenität der Zielzelle scheinen keinen Einfluß zu haben [396]. Für die zytostatische Wirkung werden unterschiedliche Mechanismen verantwortlich gemacht: so, je nach untersuchtem Tumorsystem, die Sekretion von Wasserstoffperoxid [1027], die Freisetzung von Arginase [256, 376], die Freisetzung von zytolytischen Proteasen [3], ein direkter Kontakt zwischen Makrophagen und Tumorzelle [660, 661] oder die Freisetzung von zytotoxischen C_{3a} [388] durch Spaltung des von Makrophagen selbst produzierten Komplementfaktors C_3 [1593]. Diese Spaltung von C_3 in C_{3b} und C_{3a} scheint von aktivierten Makrophagen durchgeführt zu werden. Da C_{3a} wiederum Makrophagen aktiviert, ist damit ein sich selbst aufrecht erhaltendes Aktivierungssystem gegeben [387, 388, 478, 1281].

Die Tumor-zytolytische Aktivität von Makrophagen kann auch in vivo durch Transplantationsexperimente nachgewiesen werden [1609], bei welchen Tumorzellen im Gemisch mit aktivierten Makrophagen unbehandelten Empfängertieren injiziert und das Auswachsen der Tumoren kontrolliert werden [367, 1083, 1375, 1609, 1616]. Das Ausmaß der Wachstumsinhibition ist von Tumor zu Tumor unterschiedlich [1083]. In manchen Tumormodellen werden auch keine Anhaltspunkte für eine antitumorale Wirksamkeit von Makrophagen gefunden [367]. Aktivierte zytotoxische Makrophagen können inhibitorisch wirken, sowohl auf das Wachstum von transplantierten Primärtumoren [421] als auch (nach i. v.-Injektion von Makrophagen) von spontanen Metastasen [391, 1529]. Die Behandlung mit Makrophagen-beeinträchtigenden Mitteln (Carrageenan, Silica, Kortikosteroide, Trypan blau) führt zu einem verstärkten Wachstum von Tumortransplantaten [210, 723, 754, 756]. Andererseits vermindert die Behandlung mit Immunmodulatoren, welche Makrophagen aktivieren, wie beispielsweise mit BCG [1549], Bordetella pertussis [14], Corynebacterium parvum [1170] oder Corynebacterium granulosum [975], das Anwachsen intravenös injizierter Tumorzellen, sowie, beispielsweise nach i. v.-Injektion von C. parvum, die spontane Metastasierung experimenteller Tumoren [1249] (jedoch wirken derartige Immunmodulatoren nicht ausschließlich auf Makrophagen).

Parallel hierzu kann eine Aktivierung der Makrophagen sowohl in der Peripherie als auch in der Nähe der Tumoren nachgewiesen werden [895, 1120], und abgestoßene Tumoren enthalten in beträchtlicher Anzahl aktivierte Makrophagen [870, 1242].

Eine Optimierung in der In-vivo-Aktivierung von Makrophagen besteht in der intravenösen Injektion von Immunstimulanzien (z. B. Muramyldipeptide (MDP) von BCG) oder auch Mediatoren (z. B. MAF), welche in Liposomen eingeschlossen sind. Makrophagen nehmen die Liposomen bevorzugt auf, und die in den Phagolysosomen freigesetzten Immunstimulanzien bzw. Mediatoren können die Makrophagen gezielt aktivieren [899, 1141, 1289, 1407]. Das Ergebnis bei experimentellen Tumoren war eine deutliche Verminderung des Wachstums von spontanen Metastasen [395, 399]. Da auch hier die gleichzeitige Applikation von Carrageenan oder Silica den therapeutischen Effekt der Liposomenpräparation inhibierte [399], und sowohl Carrageenan als auch Silica bevorzugt die Makrophagenfunktion hemmen [1, 17, 195, 659, 871, 1499], dürften zytotoxische Makrophagen für die antitumorale Wirkung von in Liposomen enkapsuliertem MDP oder MAF verantwortlich sein. Jedoch hemmen Carrageenan und Silica auch Lymphozyten [399, 871].

Auch die systemische Verabreichung von Makrophagen-aktivierenden Immunstimulanzien (z. B. BCG, Corynebacterium parvum, Polynucleotiden, Bestatin) führt zu einer Verzögerung oder Verminderung im Auftreten von spontanen oder chemisch induzierten Tumoren [72, 347, 456, 849, 876, 1073]. Da es durchaus möglich ist, daß zytotoxische Makrophagen die Ursache dieses Phänomens darstellen, könnten diese Zellen eine Schlüsselrolle bei der «Immune surveillance» spielen [2, 630].

Hierfür spricht auch die überdurchschnittliche Widerstandsfähigkeit von an T-Lymphozyten verarmten oder athymischen Nacktmäusen [216, 264, 470, 1188] gegen Tumorerkrankungen.

Jedoch kann diese erhöhte Resistenz gegen Tumorerkrankungen nicht nur durch die Aktivierung von Makrophagen, sondern auch durch eine Aktivierung der sogenannten natürlichen Killer-Zellen (NK-Zellen) verursacht sein [1614]. NK-Zellen sind beispielsweise in überdurchschnittlicher Menge in Nacktmäusen nachzuweisen [602].

Es ist durchaus möglich, daß alle bisherigen experimentellen Untersuchungen, die zwar in vitro und in ausgewählten In-vivo-Modellen eine deutliche Makrophagenzytotoxizität für Tumorzellen belegen, für die eigentliche In-vivo-Situation, d. h. für die autochthone Tumorerkrankung, nicht in vollem Umfang aussagefähig sind. So muß der Befund, daß die Anzahl der Makrophagen im Primärtumor mit der Verminderung der Metastasierung korreliert [331], nicht für eine kausale Beziehung zwischen diesen beiden Phänomenen sprechen [1055]. Des weiteren muß erstaunen, daß Makrophagen in einem progressiv wachsenden Tumor, gemessen an ihrer antibakteriellen Wirksamkeit, nicht aktiviert sind [1055]. Dennoch, histologische Untersuchungen an verschiedenen menschlichen Tumoren sprechen für das Bild einer Immunreaktion unter Beteiligung von Lymphozyten und Makrophagen [1464, 1465]. Zum anderen sind auch viele In-vitro-Untersuchungen zur Zytotoxizität von Makrophagen gegen Tumorzellen kritisch zu sehen: Durch Sekretion von Thymidin oder Arginase können Makrophagen entweder den Einbau von ^{3}H-Thymidin in Tumorzellen (benutzt als Parameter des Tumorzellstoffwechsels) kompetitiv inhibieren [1436] oder aber das Kulturmedium an Arginin verarmen und damit den Stoffwechsel von Tumorzellen reduzieren [256]. Beides kann einen zytotoxischen Effekt von Makrophagen vortäuschen [1055].

Dennoch, sowohl Tumor-tragende Tiere [734, 735] als auch Tumorpatienten [917] weisen eine erhöhte «Clearance rate» für i. v. injizierte Kolloide auf, was für aktivierte Makrophagen spricht. Große Tumormassen reduzieren diese «Clearance rate», und eine Exstirpation des Tumors läßt sie wieder ansteigen [917], was beispielhaft zumindest als ein Hinweis für die Existenz einer Wechselbeziehung zwischen Tumor und Makrophagen gelten dürfte.

Makrophagen sekretieren nach geeigneter Stimulierung durch Endotoxin einen Faktor, welcher in der Lage ist, Tumorzellen und Malaria-Erreger [225, 1483] zu töten. Dieser Faktor wurde Tumor-Nekrosis-

Faktor (TNF) genannt [206, 512, 513, 1078]. TNF weist mit 157 Aminosäuren Sequenzhomologien mit Lymphotoxin auf, ist von diesem jedoch unterschiedlich [7, 1118]. Der zytotoxische Wirksamkeitsmechanismus ist ähnlich wie bei Lymphotoxin noch nicht vollständig verstanden. TNF bindet an die Zielzelle über spezifische Membranrezeptoren [840, 1235] und wirkt zellzyklusspezifisch zytostatisch [269] und in dieser Hinsicht synergistisch mit Interferon [1601].

Aufgrund der zytotoxischen Aktivität von TNF auf verschiedene experimentelle und humane Tumorzellinien in vitro und der antitumoralen Wirksamkeit von TNF auf xenotransplantierte Humantumoren [553] sind klinische Studien an Tumorpatienten mit gentechnologisch hergestelltem TNF [1369] im Gange bzw. in Vorbereitung.

Natürliche Killer-(NK-) Zellen

Natürliche Killer-Zellen (NK-Zellen) sind strahlensensitiv [1106], nicht adhärent («Large granular lymphocytes» LGL) und nehmen einen Anteil von 0,5–2,5% der Lymphozytenpopulation ein [1198, 1212, 1502]. NK-Zellen können Zielzellen unabhängig von einer Immunisierung und unabhängig von der Expression von MHC-Gen-kodierten Antigenen [1447] lysieren. Aufgrund von Oberflächenmarkeruntersuchungen werden NK-Zellen weder den T- und B-Zellen noch den Makrophagen zugeordnet. Sie tragen Fc-Rezeptoren, aber ihnen fehlen Oberflächenimmunglobuline und Komplementrezeptoren [1122, 1258]. Unklar ist, ob NK-Zellen einen Funktionszustand unterschiedlicher Zell-Linien repräsentieren oder ob sie einer bestimmten Zellpopulation zuzuordnen sind [1578].

NK-Zellen sind nicht thymusabhängig und können somit in athymischen Nacktmäusen (nu/nu) und in neonatal thymektomierten Tieren in großer Zahl gefunden werden [606]. Sie tragen jedoch verschiedene T-Zell-assoziierte Marker (in der Maus Ly-1 und Ly-5 [607, 740]), die wiederum nicht T-Zell-spezifisch sind [1084, 1493]. So ist das Ly-5-Antigen auch auf Thymozyten zu finden [197, 1137]. NK-Zellen reagieren zu unterschiedlichen Anteilen mit einer Reihe von monoklonalen Antikörpern gegen T-Zell-assoziierte Antigene [346, 733, 821, 937]. NK-Zellen produzieren Interleukin II [316] und wachsen verstärkt unter Einfluß von Interleukin II [293, 1089]. Ein Teil bildet mit Schafserythrozyten Rosetten, ein anderer Teil nicht [1523].

Es wurde demzufolge vermutet, daß NK-Zellen zu einer Gruppe von Stammzellen gehören, aus denen das Thymozyten-T-Zellsystem entsteht [197]. Jedoch tragen NK-Zellen auch Oberflächenmarker, welche auf Makrophagen oder neutrophilen Granulozyten zu finden sind [161, 750, 894, 1647]. NK-Zellen scheinen entweder mit Killer-Zellen identisch zu sein (K-Zellen, verantwortlich für die Antikörper-abhängige Zell-mediierte Zytotoxizität (ADCC)), oder aber sich in der gleichen Lymphozytenfraktion anzureichern [601].

Die meßbare Aktivität von NK-Zellen bei Mäusen und Ratten, weniger beim Menschen, ist abhängig von Alter [772], des weiteren auch von dem jeweiligen Tierstamm [606, 775, 1124] bzw. dem jeweiligen HLA-Phänotyp [1258]. NK-Zellen können in unterschiedlichem Maße Tumorzellen, fetale Zellen, Virus-infizierte Zellen, aber auch bestimmte normale Zellen (unreife Thymozyten, Knochenmarkzellen und Peritonealzellen) lysieren [552, 1058, 1250, 1589], wobei die Reaktion gegen allogene Zielzellen stärker ist als gegen syngene [1058]. Da bevorzugt geringer differenzierte Zellen lysiert werden, scheinen NK-Zellen Differenzierungsantigene erkennen zu können [1251].

Im Gegensatz zu spezifischen zytotoxischen T-Zellen werden NK-Zellen nach der Zytolyse einer Zielzelle inaktiviert und können daher keine zweite Zielzelle lysieren [1523]. Es bestehen Anhaltspunkte, daß NK-Zellen Zielzellen durch die Freisetzung eines Mediators lysieren [1625]. Es ist unklar, ob es sich hierbei um eines der Lymphotoxine handelt [1625] oder aber um Proteasen [668] oder Phospholipasen [610]. Die lytische Aktivität von NK-Zellen [1454] sowie von isolierten Mediatoren [1625] kann durch einfache Zucker (D-Galactose; N-Acetyl-D-Galactosamin und -Methyl-D-Mannosid) und die der NK-Zellen durch Antikörper gegen Ganglio-N-Tetraosylceramid (Asialo-GM_1) inhibiert werden [1640]. Asialo-GM_1 wird auch von Makrophagen exprimiert [610].

NK-Zellen scheinen im Gegensatz zu Makrophagen nicht Sauerstoffradikale bilden zu können [484]. NK-Zellaktivität kann in Mäusen durch Virusinfektionen erhöht werden, des weiteren durch Gabe von Immunstimulanzien wie beispielsweise BCG, Corynebacterium parvum oder Bestatin [138, 604, 605, 1355, 1613] oder nach Applikation von Interferon-Inducern [1063] oder von Interferon selbst [311, 312, 468, 601, 609, 1197], des weiteren auch nach Injektion von Tumorzellen [276, 599–601]. Die Aktivierung von NK-Zellen durch Interferon verdient besondere Betrachtung: Zum einen kann sie auch beim Menschen beobachtet werden [1197], und zwar mit α-, β- und γ-Interferon [601, 1522], zum anderen ist sie durch Antikörper gegen Interferon zu inhibieren [118, 1064]. An separierten «Large granular lymphocytes» konnte gezeigt werden, daß Interferon direkt, d. h. ohne zusätzliche Hilfe durch andere Zellen die NK-Zellaktivität zu steigern vermag [1502], und zwar möglicherweise durch Zunahme der Rezeptoren auf den NK-Zellen für die jeweilige Zielzelle und/oder durch Wandlung von inaktiven NK-Zellen in zytolytische NK-Zellen [601]. Dieser Einfluß von Interferon wird innerhalb einer kurzen Expositionszeit von ca. 1 h bereits meßbar [1523], so daß durchaus Interferon als einer der verantwortlichen Mediatoren für die Aufrechterhaltung eines bestimmten Grades an NK-Zellaktivität verantwortlich gemacht werden kann [601]. Da NK-Zellen selbst nach Stimulation durch Viren, mitogenen Immunstimulanzien oder Tumorzellen Interferon sekretieren [313, 601, 1252, 1519, 1523], besteht hier eine gewisse Autonomie im Sinne der Aktivierung von zytolytischen NK-Zellen, zum anderen eine Einflußmöglichkeit auf die Makrophagenaktivierung und Antigen-spezifische humorale und zelluläre Immunreaktionen (siehe S. 30, 34).

Ob Umwelteinflüsse den exogenen Reiz für einen über Interferonfreisetzung aufrechterhaltenen Grad an NK-Zellaktivität spielen [224, 1197], ist fraglich, da auch keimfreie Tiere eine deutliche NK-Zellaktivität aufweisen [781]. Alle drei Typen (α, β, γ) von Interferon sind in der Lage, bestimmte Zielzellen (Zellen mit hoher Sensitivität für Interferon) gegen NK-Zellen widerstandsfähig zu machen [1519, 1523, 1524]. Dieser Schutz durch Interferon ist dosisabhängig und reversibel. Er beeinflußt nicht die Zytotoxizität von K-Zellen durch ADCC oder die Antikörperbedingte Komplement-abhängige Zytolyse [1523]. Der Schutz der Zielzelle durch Interferon wird durch Virusinfektion oder durch Inhibition der RNA-Synthese verhindert [1524]. Des weiteren sind die meisten Tumorzellen nicht durch Interferon gegen NK-Zellen zu schützen [1523]. Interferon scheint demnach sowohl die Stärke der NK-Zellaktivität zu regulieren als auch gesunde, normale Zellen vor dem Angriff der NK-Zellen zu schützen [1259, 1523].

Ein Großteil derjenigen Substanzen, welche die NK-Zellen aktivieren, setzt auch Interferon frei. Dennoch scheint Interferon nicht der einzige positive Regulationsfaktor für NK-Zellen zu sein, da beispielsweise auch polyklonale oder monoklonale Antikörper eine Aktivierung der NK-Zellen bewirken können [179, 180].

Inhibiert wird die NK-Zellaktivität durch Prostaglandine [311,

312, 324, 1512, 1513]. Diese Inhibition ist durch Endomethacin hemmbar und des weiteren reversibel [178]. Da Makrophagen bekanntermaßen Prostaglandine sekretieren, könnte die nach Behandlung mit den verschiedensten Immunstimulanzien überraschenderweise auch in vivo auftretende Suppression der NK-Zellaktivität durch aktivierte Makrophagen verursacht worden sein [249, 901, 1260]. Ähnliches dürfte für die erniedrigte NK-Zellaktivität Neugeborener zutreffen [249].

Tumoren können sowohl in experimentellen Systemen als auch beim Menschen zu einer Suppression der NK-Zellaktivität führen. Durch Entfernung der adhärenten oder phagozytierten Zellen kann diese Suppression aufgehoben werden, was für Makrophagen als Ursache dieser Inhibition spricht [360, 462, 463, 1566]. Weitere Faktoren, die NK-Zellen inhibieren können, sind bestimmte Tumorpromotoren wie beispielsweise Phorbolester [484, 755].

Ähnlich wie bei den Makrophagen wird vermutet, daß NK-Zellen im Rahmen der Tumorabwehr eine entscheidende Rolle spielen [601, 1212]. Unterstützt wird diese Vermutung durch verschiedene experimentelle Untersuchungen:

So kann die Injektion von isolierten NK-Zellen (im Gegensatz zu Milzzellen) das Wachstum von Tumortransplantaten in sublethal bestrahlten Mäusen total inhibieren [197]. NK-sensitive Tumortransplantate wachsen in Mäusen mit genetisch bedingter hoher NK-Zellaktivität schlechter als in Mäusen mit niedriger NK-Zellaktivität [556, 772, 1345]. Thymuslose Nacktmäuse oder T-Zell-defiziente F_2-Hybride weisen in vitro eine erhöhte NK-Zellaktivität auf und sind widerstandsfähiger gegen NK-sensitive syngene Tumortransplantationen [774, 1124]. Nach Injektion von T-Zell-verarmtem Knochenmark von Spendern mit hoher oder niedriger NK-Zellaktivität in thymektomierte und bestrahlte F_2-Empfänger zeigen die Empfänger in gleicher Weise Schutz vor dem Wachsen von semisyngenen Tumortransplantaten wie die Spender [537, 1203]. NK-Zellen können aus experimentellen Primärtumoren isoliert werden [462], jedoch nicht, wenn die Tumormasse zu groß ist und auch nicht aus klinischen Tumorproben [610]. Gealterte Mäuse weisen im allgemeinen eine erhöhte Rate spontaner Tumoren auf und sind gleichzeitig weniger gut als junge Mäuse in der Lage, NK-Zell-sensitive Tumortransplantate abzustoßen [537]. Eine künstlich erhöhte NK-Zellaktivität in Mäusen (z. B. bewirkt durch Verarmung an B-Zellen nach chronischer Behandlung mit Antikörpern gegen die schwere Kette von IgM) geht einher mit einer erhöhten Widerstandsfähigkeit gegen eine Tumorinduktion durch das Moloneysarkoma-Virus, beeinträchtigt das Wachstum von Tumortransplantaten und hemmt die spontane Metastasierung [165]. Metastasen eines spontanen Mammatumors in der Ratte weisen eine deutlich geringere Empfindlichkeit für NK-Zellen auf als die Zellen des Primärtumors, jedoch ist die Resistenz gegen NK-Zellen von der Organlokalisation der Metastase abhängig [171].

Des weiteren sind die «beige»-Mäuse (bg/bg), die für NK-Zellen selektiv defizient sind, ansonsten jedoch ein normales Immunsystem aufweisen, für Wachstum und Metastasierung von Tumortransplantaten stärker empfänglich als die phänotypisch normalen Heterozygoten (+/bg) [739, 1477]. Werden jedoch NK-Zell-resistente Tumoren transplantiert, ist kein Unterschied im Tumorwachstum zwischen bg/bg und +/bg zu sehen [1477]. Auch die Tumorinduktion durch Karzinogene ist höher in bg/bg als in +/bg [1212]. Eine Injektion von NK-Zellen in beige-Mäuse (bg/bg) verstärkt in diesen Tieren die HvG-Reaktion gegen Knochenmarktransplantate, schützt sie vor dem Anwachsen nachfolgend i. v. injizierter syngener Tumorzellen und verhindert die Entstehung von spontanen Leukämien, induziert durch Bestrahlung vor der Applikation der NK-Zellen [1579]. In beige-Mäusen ist des weiteren das Auftreten von Lymphomen deutlich vermehrt [902].

Diese experimentellen Modelle geben insgesamt durchaus Anhaltspunkte, daß NK-Zellen im Rahmen der Tumorabwehr eine Rolle spielen können. Da jedoch keine «reinen» Modelle verfügbar sind, in welchen alle anderen auch für die Tumorabwehr in Frage kommenden Zellsysteme ausgeschaltet werden können, ist der Beweis für die dominierende Rolle der NK-Zellen bislang nicht möglich [608]. Speziell auch dieser Tatbestand macht die Prüfung der Bedeutung von NK-Zellen in der Klinik schwierig.

In Form des Chediak-Higashi-Syndroms wurde eine genetisch bedingte Abnormalität gefunden, die den beige-Mäusen (bg/bg) entspricht [1212]. Diese Patienten verfügen, ähnlich wie die beige-Mäuse, über ein intaktes restliches Immunsystem [521, 1213]. Aufgrund der bisherigen Erfahrungen endet das Chediak-Higashi-Syndrom bei der Mehrzahl der Patienten in einer lymphoproliferativen Erkrankung [119, 299].

Untersuchungen der NK-Zellaktivität an Patienten mit unterschiedlichen Tumoren zeigen eine gewisse Korrelation zum Tumorstadium und zur Behandlung [620, 621, 1220, 1555, 1646]. Jedoch können aussagefähige Korrelationen zur Prognose von unterschiedlichen Tumorerkrankungen anscheinend nur dann gefunden werden, wenn die NK-Zellaktivität gegen autologe Tumorzellen ermittelt wird [805]. Trotz deutlicher Aktivität gegen allogene Tumorzellen kann jedoch die Zytotoxizität gegen autologe Tumorzellen vollkommen fehlen [1555].

Granulozyten

Die Rolle von Granulozyten bei der Tumorabwehr ist bislang noch nicht eindeutig geklärt worden. Es bestehen jedoch Anhaltspunkte, daß Granulozyten direkt zytotoxisch auf Tumorzellen wirken können. So werden Kulturzellen durch Granulozyten von ihrer Unterlage abgelöst [408, 827, 908]. Aktivierte Granulozyten können Tumorzellen lysieren, sowohl in vitro [1126] als auch in vivo nach i. p.-Tumortransplantation [328]. Auch nicht-aktivierte Granulozyten scheinen einen zytolytischen Effekt auf Tumorzellen zu haben [215]. Zu diesem Befund gibt es jedoch auch widersprechende Untersuchungen [211, 827, 1473]:

So konnte keine direkte zytotoxische, jedoch eine zytostatische Wirkung von Granulozyten, bevorzugt auf Tumorzellen und Virus-infizierte Zellen, weniger auf Normalzellen nachgewiesen werden [827]. Der Wirkungsmechanismus, welcher der Zytostase zugrunde liegt, ist unbekannt, jedoch scheinen Sauerstoffradikale, speziell Wasserstoffperoxid, eine Rolle zu spielen, da ein Teil der zytostatischen oder zytotoxischen Wirkungen durch Katalase zu hemmen ist [223, 827].

Die Bedeutung der größtenteils in vitro ermittelten zytostatischen oder zytotoxischen Effekte von Granulozyten für die Tumorabwehr ist unklar. Granulozyten sind häufig am oder im Tumorgewebe zu finden, möglicherweise bedingt durch Chemotaxis auf TAA oder Komplementspaltprodukte [447, 1070].

Antikörperabhängige, zelluläre Zytotoxizität
(Antibody-dependent cellular cytotoxicity = ADCC)

Voraussetzung für eine ADCC sind Antikörper, die spezifisch an eine Zielzelle binden und zytotoxische Zellen, welche über ihre Fc-Rezeptoren mit den Fc-Teilen der gebundenen Antikörper reagieren und diese Reaktion mit Zytolyse der Zielzelle beantworten [904, 914, 993, 1119]. Die beteiligten Antikörper gehören vorzugsweise zur IgG-Klasse, der Subtyp richtet sich nach der Spezifität des Fc-Rezeptors der beteiligten zytotoxischen Zelle. Lymphozyten, aber auch Makrophagen, neutrophile und eosinophile Granulozyten und Thrombozyten sind zur ADCC fähig [200, 904]. ADCC-Lymphozyten, auch «Killer»-Zellen (K-Zellen) genannt, gehören zur Klasse der «Null»-Zellen, da sie weder die herkömmlichen Oberflächenmarker der B-Zellen noch der T-Zellen tragen [1172]. Aufgrund ihrer Oberflächenmarker und funktionellen und morphologischen Eigenschaften werden sie den natürlichen Killer-Zellen zugeordnet [601]. Jedoch scheinen Unterschiede in der Trypsinsensitivität (Fc-Rezeptor von K-Zellen ist Trypsin-resistent) und in der Bindungsaffinität der Fc-Rezeptoren zu IgG-Aggregaten zu bestehen [124, 749, 828]. In diesem Zusammenhang besitzen die Fc-Rezeptoren von Monozyten die stärkste Affinität [1144] und K-Zellen eine größere als NK-Zellen. Unter Zuhilfenahme von Glutaraldehyd-stabilisierten Zielzellen und unter Berücksichtigung der Avidität der Fc-Rezeptoren konnten K-Zellen von NK-Zellen getrennt und gesondert angereichert werden [1036]. Jedoch liegen auch dem widersprechende Untersuchungsergebnisse vor [288]. Die relativ hohe Avidität der Fc-Rezeptoren von K-Zellen scheint aus der hohen Dichte der Fc-Rezeptoren auf K-Zellen zu resultieren [429, 910]. Es ist nicht auszuschließen, daß diese Unterschiede zwischen K-Zellen und NK-Zellen einem Reifungsprozeß entspringen [1036].

Der Mechanismus, welcher zur Zytolyse durch K-Zellen im Rahmen der ADCC führt, ist unbekannt. Es gibt Anhaltspunkte, daß Komplement-Komponenten [1460] oder Phospholipasen [433] an der Zytolyse durch ADCC beteiligt sein könnten. Ähnlich wie bei NK-Zellen steigt die Aktivität der K-Zellen nach Applikation von Interferon-Inducern oder nach Interferongabe an [609, 1197], können K-Zellen Interferon produzieren [1523, 1524] und werden K-Zellen nach der Interaktion mit der Zielzelle inaktiviert und können eine zweite Zielzelle nicht lysieren [1656]. Im Gegensatz zu der Situation bei der NK-Zelle wird jedoch die Zielzelle durch Interferon nicht gegen die ADCC von K-Zellen geschützt [1524].

Über die klinische Relevanz der ADCC in Hinblick auf Tumoren gibt es nur beschränkte und widersprüchliche Informationen [945, 1474]. Sie alle werden durch falsch-positive Befunde bei Ermittlung der ADCC beeinträchtigt, bedingt durch kreuzreagierende Antikörper oder sogar IgG-Aggregate, welche in geringsten Mengen durch Bindung der K-Zelle an die Zielzelle eine ADCC auslösen können [1181]. Ebenso wie

K-Zellen können Makrophagen nach Aktivierung [630, 1632] eine ADCC durchführen, wobei die zytolytischen Mechanismen den bereits für die direkte Zytotoxizität beschriebenen (siehe S. 30) gleichen dürften. So ist die ADCC von Makrophagen durch Thioglykolat, einem Inhibitor von Wasserstoffperoxid, zu inhibieren [2,3 1027]. Bemerkenswert ist jedoch, daß die ADCC durch Makrophagen auch durch Zugabe geringer Mengen von LDL und VLDL (Low-density-Lipoproteine bzw. Very-low-density-Lipoproteine, gewonnen aus Tumoraszites) inhibiert werden kann [630, 1633]. Die Zusammensetzung der Membranlipide scheint demnach für die Zytotoxizität entscheidend zu sein. Ob eine ADCC von Makrophagen gegen Tumorzellen besteht, wird kontrovers diskutiert [4,5]. Es gibt Anhaltspunkte, daß nicht alle Makrophagen zur ADCC fähig sind, beziehungsweise, daß es innerhalb der Makrophagen unterschiedliche Formen der ADCC geben mag [4].

Für eine ADCC besitzen humane Makrophagen Fc-Rezeptoren für Immunglobuline des IgG_{2a}/IgG_1-Isotyps (FcI-Rezeptoren) und des IgG_{2b}-Isotyps (FcII-Rezeptoren) (Übersicht bei [4]). Diese reagieren nicht nur mit isologen Immunglobulinen, sondern unter bestimmten Bedingungen (10 bis 20 Tage in Kultur gehalten) auch mit monoklonalen Antikörpern der Maus [1445]. So konnte eine ADCC von humanen Monozyten mit Immunglobulin G_{2a} (nicht aber mit anderen Subklassen) von der Maus spezifisch für TAA humaner Kolonkarzinomzellen erzielt werden [1445], was neue Wege einer Therapie mit monoklonalen Antikörpern eröffnet [4]. In Ergänzung zu K-Zellen und Makrophagen können auch neutrophile Granulozyten eine ADCC gegen Tumorzellen bewirken [444, 827]. Auch hier scheint der zytotoxische Mechanismus ähnlich zu sein, wie er bereits für die direkte Zytostase beschrieben wurde. Erstaunlicherweise sind auch Thrombozyten in der Lage, eine ADCC gegen Tumorzellen auszuführen [903, 1363].

Beeinträchtigung der Immunabwehr

Wir sind mit der Tatsache konfrontiert, daß trotz der verschiedensten bislang bekannten antigenspezifischen sowie unspezifischen immunologischen Abwehrmechanismen, welche dem Organismus zur Verfügung stehen, gutartige wie bösartige, antigene und nichtantigene Tumoren entstehen und progressiv wachsen. Wenn diese Abwehrmechanismen wirklich die ihnen zugeschriebene Schlüsselrolle bei der Überwachung der Tumoren besitzen, muß demnach gefolgert werden, daß Tumorzellen über die Fähigkeit verfügen, sich einem immunologischen Angriff entziehen zu können. Zum anderen dürften auch Fehlsteuerungen und Mängel im Abwehrsystem des Wirtes das Wachstum von Tumoren begünstigen (Übersicht bei [842, 1201]). In diesem Zusammenhang ist das Phänomen des «Sneaking through» [674, 1074] zu erwähnen, bei welchem nach In-vivo-Transplantation kleine Mengen von antigenen Tumorzellen progressiv wachsen, mittlere Mengen abgestoßen werden und größere Mengen wiederum wachsen. Dieses «Sneaking through» wird möglicherweise durch die Zelldosis-abhängige Stimulation von T-Suppressorzellen bewirkt [222, 817].

Eine Einsicht in die Einzelheiten der immunologischen Wechselbeziehungen zwischen Tumor und Wirt ist derzeit nur in Ansätzen vorhanden. Sie müssen anderen, möglicherweise gleich bedeutsamen, jedoch auch noch nicht vollständig geklärten Möglichkeiten der gegenseitigen Einflußnahme zugeordnet werden, so beispielsweise biochemische Veränderungen am Ort des Tumorwachstums und Vaskularisation und Blutversorgung des Tumors.

Quantitative oder qualitative Mängel des Immunsystems

Voraussetzung für eine Tumorabwehr ist die Immunkompetenz des Organismus. Eine Beeinträchtigung dieser Immunabwehr ist durch Alter (Übersicht bei [751]) und durch Erkrankungen des Immunsystems gegeben [769] oder tritt im Gefolge einer immunsuppressiven Therapie oder Manipulation auf, beispielsweise im Rahmen der Chemotherapie von Organabstoßungen oder Tumoren [18, 507, 834, 1117, 1277]. Diese Beeinträchtigung der Immunabwehr kann, soweit erfaßbar, von einem verstärkten Auftreten von Tumoren begleitet sein, nach Chemotherapie besonders von Tumoren des lymphoretikulären Gewebes [1068], jedoch auch von soliden Tumoren unterschiedlichster Art [333, 1277]. Als Ursache für diese lymphatischen Tumoren werden neben der kanzerogenen Potenz einiger dieser Therapeutika die bevorzugte Elimination der T-Suppressorzellen [48, 461], chronische Stimulation durch Antigene bei Organtransplantationen [31] oder Aktivierung tumorassoziierter Viren [110, 636] angesehen.

Durch den Tumor selbst bedingte systemische Beeinträchtigungen der allgemeinen Immunabwehr treten im Regelfall nur bei fortgeschrittenen Tumorerkrankungen auf und hier besonders bei denjenigen, welche die lymphatischen und blutbildenden Organe betreffen. Zusätzlich gibt es Befunde, die bei Mäusen auf eine durch Tumoren bedingte unspezifische Suppression der T-Zellen in Thymus und Lymphknoten hinweisen [900, 1233].

Da athymische Nacktmäuse oder neonatal thymektomierte Tiere nicht häufiger an Tumoren erkranken als die jeweiligen Normaltiere bzw. normalen Hybriden, dürfte die Suppression des T-Zellsystems jedoch nicht die einzige Ursache für die Tumorentstehung sein [1244].

Mangelnde Tumorerkennung

Derzeit werden verschiedene Faktoren diskutiert, mit Hilfe derer sich Tumoren der Erkennung und Abwehr durch die spezifische Immunabwehr entziehen können.

Auffallend ist, daß die Malignität einer Tumorzelle umgekehrt proportional ihrer Immunogenität bzw. Antigenität zu sein scheint (Übersicht bei [1272]). So sind metastasierende Tumoren im allgemeinen nicht oder nur schwach immunogen [1272]. Weiterhin ist auffallend, daß im Gegensatz zu chemisch oder Virus-induzierten Tumoren ein Großteil der spontanen experimentellen Tumoren nicht immunogen ist [624, 973]. Das mag in der a priori mangelnden Ausbildung von Tumorantigenen einen Grund haben [414, 624, 799, 800, 1588] oder aber in einer durch die Immunabwehr bewirkten Selektion von nicht-immunogenen, Tumorantigen-negativen und damit immunresistenten Tumorzellen aus einer heterogenen Zellpopulation. Hinweise hierfür bestehen bei verschiedenen experimentellen Tumoren [386, 393, 394, 398, 689, 1488].

Des weiteren scheint auch bei Humantumoren die Dichte der Tumorantigene auf der Zellmembran von der Wachstums- bzw. Zellteilungsphase [1039] und von dem Umgebungsmilieu [142] abhängig zu sein. Innerhalb eines Tumors sowie zwischen Primärtumor und Metastasen bzw. zwischen den Metastasen können die jeweiligen Tumorzellen unterschiedliche Tumorantigene [375, 411, 1025, 1272, 1457, 1617, 1636] und auch Unterschiede in der Immunogenität aufweisen [287, 778]. Tumorzellvarianten innerhalb einer Tumorzellpopulation, die ihre Antigenität verloren haben, konnten durchaus in einer Reihe von Tumorsystemen nachgewiesen werden [107, 397, 1585, 1636] und hier besonders gehäuft bei Metastasen [295].

Ein anderer Mechanismus besteht in der Modulation von Tumormembranantigenen, ausgelöst durch die spezifische T-Zell-mediierte humorale und/oder zelluläre Immunreaktion. Diese Modulation führt entweder zu einer phänotypischen reversiblen [151, 1077] oder zu einer genetisch stabilen Veränderung [1272] im Antigenmuster der Tumorzellen speziell auch der entstehenden Metastase, so daß diese Tumorzellen gegen die spezifische Immunreaktion resistent werden. Am Beispiel des metastasierenden EsB-Lymphoms konnte gezeigt werden, daß diese Resistenzentwicklung, welche mit mangelnder Zytolyse durch spezifische CTL einherging, in T-Zell-defizienten Nacktmäusen (nu/nu) nicht auftrat und demnach ein T-Zell-abhängiger, in seinem biochemischen Ablauf noch ungeklärter Prozeß ist [135, 1272].

Da für die zelluläre Erkennung von Tumorantigenen im efferenten und möglicherweise auch im afferenten Weg die gleichzeitige Exprimierung von MHC-Genprodukten entscheidend sein kann (siehe S. 26, 30), besteht die Möglichkeit, daß quantitative [280, 569, 1279, 1503, 1504] oder qualitative Veränderungen beispielsweise durch Auftreten fremder, haplotypischer Antigene [451, 655, 967, 1278] Tumorzellen nicht erkennbar oder beispielsweise für CTL unangreifbar machen.

So wurde in verschiedenen experimentellen Tumorsystemen eine Variabilität in MHC-Genprodukten [925] nachgewiesen, und es wurde in der Tat gefunden, daß Tumorzellen aus metastasierenden Klonen eines T_{10}-Sarkoms andere Alloantigene exprimieren als die nicht-metastasierenden Klone [747] und daß Veränderungen in der Metastasierungsrate mit veränderten MHC-Genprodukten einhergingen [345, 747, 748].

Des weiteren bestehen Anhaltspunkte, daß bestimmte Adenoviren onkogen werden können, indem sie in der infizierten Zelle zu einer verminderten Expression von Transplantationsantigenen führen und dadurch die transformierte Zelle für die T-Zell-Abwehr unangreifbar machen [98]. Auch die unspezifische Immunreaktion gegen Tumoren ist durch den Zellmembranunterschied zwischen Primärtumor und Metastase betroffen: So lysieren NK-Zellen deutlich weniger Tumorzellen aus Metastasen als aus dem Primärtumor [171].

Ein weiterer, in seiner Bedeutung noch nicht abschätzbarer Mechanismus, der Immunantwort zu entgehen und an Malignität zu gewinnen, ist die spontane Fusion von Tumorzellen mit Makrophagen [855] oder B-Lymphozyten [281] oder anderen Zellen von Normalgeweben [483, 764]. Derartige Fusionen konnten bei experimentellen Tumoren in syngenen Systemen, des weiteren sowohl bei Humantumoren mit Zellen des Empfängertiers [483] als auch mit homologen Zellen [34] nachgewiesen werden.

Abscheidung von Tumorantigenen

Tumoren können Tumorantigene von der Zellmembran abstoßen [1048, 1444, 1605]. Das Ausmaß dieser Freisetzung ist von Tumorantigen zu Tumorantigen unterschiedlich. Es scheint von dem Glykosylierungsgrad des jeweiligen Antigens beeinflußt zu sein und ist bei Antigenen ohne Kohlenhydrate niedrig [1026]. Des weiteren führt Interferon oder Hyperthermiebehandlung zu einer Verstärkung dieser Freisetzung [1039]. Abgesonderte Tumorantigene können durch Bindung an die jeweiligen spezifischen Antigenrezeptoren Effektorzellen neutralisieren [52–54, 261] und dadurch zu einer lokalen oder auch peripheren Hemmung beispielsweise der CTL führen [52] oder aber nach Bindung an Antikörper (Immunkomplexbildung) die komplementabhängige Zytolyse von Tumorzellen verhindern. Zum anderen können diese Immunkomplexe immunsuppressiv wirken (siehe S. 57). Zusätzlich können Tumorantigene T-Helfer-Zellen und damit eine Immunantwort (siehe S. 26) oder aber auch T-Suppressorzellen stimulieren und damit eine Blockade der Immunantwort (siehe S. 55) verursachen. Ob das eine oder das andere Phänomen induziert wird, kann von der Quantität und/oder der Qualität des Antigens abhängen. So sind voneinander unterschiedliche Antigene eines insgesamt nicht immunogenen Tumors beschrieben worden, die das Immunsystem entweder spezifisch stimulieren oder supprimieren. Beide Antigene kommen gleichzeitig im Tumor vor, wobei jedoch die supprimierenden Antigene dominieren können [804, 1024, 1116]. Tumoren mit hoher Absonderungsrate von Tumorantigenen zeigen häufig auch eine hohe Metastasierungsrate [10, 262, 780, 1451].

Inhibition von Makrophagen

Es gibt Anzeichen für eine Unterdrückung unspezifischer Immunreaktionen durch Tumoren: speziell Makrophagen können in allen ihren Funktionen lokal und systemisch gehemmt werden. Beispiele hierfür sind die Hemmung von Entzündungsreaktionen im Tumorgewebe nach Provokation mit Reizmitteln [919] und die Begünstigung des intratumoralen Bakterienwachstums [1414], möglicherweise dadurch, daß Makrophagen in den Tumoren keine Aktivierung zur Bakterizidie erfahren [1414].

Weitere Beispiele sind die systemische Hemmung von Makrophagen in ihrer Reaktion auf Entzündungsreize, wie am Modell der Makrophagen-Migration in der Bauchhöhle bei der Maus [962, 1403], der Ratte [331, 332, 1050] und am Meerschweinchen nachgewiesen [99], wie mit Hilfe der Hautfenstertechnik beim Menschen gezeigt und wie auch mit Hilfe von Chemotaxisversuchen in vitro bei der Maus und beim Menschen bestätigt werden konnte [123, 565, 962, 1033, 1050, 1234, 1401, 1403].

Untersuchungen am Menschen deuten darauf hin, daß besonders Tumorpatienten mit schlechter Prognose eine Suppression der Makrophagenmigration aufweisen [1234], andererseits eine Tumorresektion [z. B. die Ablatio mammae beim Brustkrebs) zu einer Wiederherstellung des Normalzustandes führt [1404]. Syngenes und allogenes Normalgewebe hat keinen Einfluß auf Makrophagen [1403]. Die Bedeutung dieser Ergebnisse zur Chemotaxis wird jedoch durch gegenteilige Befunde eingeschränkt [1091]. Im Gegensatz zu Makrophagen scheint die Funktion von Granulozyten durch Tumoren nicht beeinträchtigt zu werden [1050, 1403].

Tumoren beeinträchtigen des weiteren die Phagozytose von Makrophagen, gemessen an der «Clearance»-Rate kolloidaler Kohle [1091] und an der Resistenz gegen bakterielle Infektionen [1052]. Diese Resistenzminderung nach Injektion von Tumorzellen ist meist vorübergehend und wird gefolgt von einer überdurchschnittlichen Resistenzsteigerung, die mit einer gleichzeitigen Widerstandsfähigkeit gegen eine weitere Tumorzellinjektion einhergeht [1053]. Diese Inhibition von Makrophagen mag durch Faktoren verursacht sein, die von der Tumorzelle selbst, gleich welchen Typs und welcher Herkunft [1033, 1034], produziert werden und gegebenenfalls nur lokal wirken [218, 377, 1054, 1130, 1202].

In-vitro-Untersuchungen mit unterschiedlich gereinigten Präparationen aus verschiedenen Tumorzellen geben Hinweise auf hitzestabile Faktoren [1202] mit Molekulargewichten zwischen 1000 und 10 000 Dalton, welche in der Lage sind, die Migration [1033, 1091], die Bakterizidie [1052], die Zellspreitung [378] und die Zytotoxizität [218] von Makrophagen zu inhibieren. Zwar können progressiv wachsende Tumoren viele Makrophagen enthalten, diese Makrophagen sind jedoch geringer zytotoxisch als solche aus sich zurückbildenden Tumoren [1242, 1481]. Einschränkend ist zu sagen, daß die Versuche zur Inhibition der Zytotoxizität von Makrophagen durch Tumorpräparate keine einheitlichen Ergebnisse aufweisen [1201]. So existieren auch Befunde über Unwirksamkeit [1217] oder sogar über eine Steigerung der Zytotoxizität [699]. In vivo kann die gemeinsame Injektion von immunsuppressiven Faktoren aus Tumoren und Tumorzellen zu einer Verstärkung des Tumorwachstums [1130] oder zu einer Beschleunigung im Anwachsen der Tumoren führen [1033]. Die Inaktivierung von Makrophagen durch Tumoren bzw. Tumorprodukte mag nur ein vorübergehender Effekt sein, der von einer Phase der Aktivierung gefolgt wird [630, 1201]. Die systemische oder lokale Inaktivierung von Makrophagen [1414] kann jedoch zur Folge haben, daß sich Tumoren bilden, die nachfolgend aktivierte Makrophagen nicht mehr beseitigen können [1201]. Gegen diese Annahme spricht jedoch der Befund, daß zumindest bei einigen [1049], aber nicht bei allen Tumoren [1130, 1402] der Tumor eine gewisse Größe erreicht haben muß, um auf Makrophagen immunsuppressiv zu wirken.

Aktivierung von Suppressorzellen

Tumoren können antigenspezifisch T-Suppressorzellen stimulieren [166, 492, 784, 1023, 1267, 1515] und hierdurch die zelluläre Immunreaktion inhibieren [785], was ein verstärktes Tumorwachstum ermöglicht [167, 436, 437].

T-Suppressorzellen aus dem Thymus und der Milz von tumortragenden Tieren, injiziert in spezifisch immunisierte Tiere, reduzieren eine bestehende Immunität gegen Tumoren [436, 437, 515]. Die supprimierende Aktivität kann durch Anti-T-Antikörper eliminiert werden [516, 1515, 1536]; sie kann andererseits auch durch lösliche Faktoren übertragen werden, isoliert von Thymuszellen oder Milzzellen von Tumorträgern. Die löslichen Faktoren besitzen Antigenstrukturen, kodiert von Genen der MHC-Region, und Antigenstrukturen der jeweiligen Tumoren [168, 169, 1121]. Antikörper gegen diese MHC-Gen-kodierten Antigenstrukturen inhibieren die supprimierende Wirkung von Suppressor-Faktoren auf die Immunantwort und führen zur Regression von Tumoren [515, 516].

Die Bildung von Suppressorzellen kann beispielsweise durch Thymektomie reduziert und zum Beispiel durch Adjuvanzien wie «Komplettes Freundsches Adjuvans» [1193] oder durch eine kleine, nicht immunogen wirkende Anzahl von Tumorzellen [222, 817] stimuliert werden. Das seit längerem bekannte progressive Wachstum kleiner Tumortransplantate im Vergleich zu großen Inokula [1074, 1162] mag somit in der Induktion von spezifischen Suppressorzellen seinen Grund haben [1023].

Des weiteren bestehen Anhaltspunkte dafür, daß neben immunogenen gleichzeitig auch hiervon unterschiedliche suppressogene antigene Determinanten auf der Tumorzellmembran existieren können [1024]. Diese suppressogenen Determinanten besitzen häufig ein größeres Molekulargewicht als die immunogenen Strukturen [804, 1640] und sind in ihrer suppressogenen Wirkung dominant über die immunogenen Membrankomponenten [6, 700, 804, 1302].

Tumorspezifische Suppressorzellen können im Rahmen der Karzinogenese entstehen, beispielsweise durch ultraviolette Bestrahlung [278, 406, 407, 1412]. In Konsequenz dürfte bei experimentellen Tumoren je nach Behandlung entweder ein Rückgang oder aber eine Progression des Tumorwachstums zu beobachten sein. Hierbei können T-Suppressorzellen die spezifische Imunantwort sowie die Antikörperantwort [168, 169] und die CTL [639] reduzieren.

Das Auftreten von spezifischen Suppressorzellen kann mit der Entstehung von Metastasen korrelieren [1642]. Des weiteren sind von menschlichen [23, 1195, 1594] sowie von experimentellen Tumoren [519, 732, 841, 1135, 1509, 1615] die Bildung von unspezifischen Inhibitionsfaktoren für T-Lymphozyten beschrieben worden, gemessen beispielsweise an der Unterdrückung der DTH-Reaktion [332, 1033].

Des weiteren können T-Suppressorzellen Suppressorfaktoren absondern, die an Makrophagen binden, welche wiederum durch Freisetzung eines unspezifischen Inhibitors für immune T-Zellen zu einer Inhibition der Kontakt-Sensitivität führen [33, 818, 1174, 1653]. Suppressorzellen können außerdem auch NK-Zellen hemmen [1261].

Zusätzlich gibt es Anhaltspunkte für die Existenz von Suppressorzellen, welche nicht zu den T-Zellen gehören. So können, wie bereits beschrieben (siehe S. 26, 53) Makrophagen über die Sekretion von Prostaglandinen nicht antigenspezifisch immunsuppressiv wirken und beispielsweise zu einer Verschlechterung der DTH-Reaktion bei Tumorpatienten führen [168, 169, 491]. Zusätzlich scheinen Tumorpatienten immunsuppressive Faktoren zu besitzen, die mit B-Zellen interagieren, welche wiederum Suppressorzellen aktivieren können [1045] oder selbst

Suppressorzell-Eigenschaften besitzen [493]. So ist in Diskussion, ob Suppressorzellen, die weder zu T-Zellen noch zu Makrophagen gehören, beim multiplen Myelom vorkommen und hier entscheidend an der häufig bei dieser Krankheit auftretenden Immunsuppression teilhaben [1096].

Hemmung durch Antikörper

Antikörper können direkt das Tumorwachstum begünstigen [20, 115]. Dieser Effekt scheint von zellvermittelten Immunreaktionen unabhängig zu sein [108, 403]. Von größerer Bedeutung mögen jedoch diejenigen Antikörper sein, welche an Tumorantigene binden und dadurch Erkennung und zelluläre Immunabwehr durch Lymphozyten und Makrophagen behindern [383, 1561]. Dieses als «Immunologische Interferenz» bezeichnete Phänomen [235] konnte auch bei Tumortransplantationsexperimenten beobachtet werden, jedoch hier in allogenen Systemen [731]. Besonders Spaltprodukte von Antikörpern, die aufgrund fehlender Fc-Fragmente weder Komplement aktivieren noch eine ADCC vermitteln können, sind zu dieser Blockade in der Lage. In syngenen oder autochthonen Tumormodellen scheinen sie jedoch keine größere Bedeutung zu haben [255].

Eine andere Auswirkung von Antikörpern ist die Antigenmodulation. Sie beinhaltet eine durch Bindung des spezifischen Antikörpers verursachte, meist vorübergehende Veränderung in der Exprimierung von Tumorantigenen, beispielsweise durch Abstoßung und/oder durch Internalisation. Antigenmodulation konnte beim Thymus-Leukämieantigen [152] oder beim Mammatumorvirus-Antigen [633, 1435] bzw. bei Mammatumoren [1048] nachgewiesen werden. Die Antigenmodulation bewirkt eine Resistenz gegen Antikörper-mediierte, Komplementabhängige Zytolyse. Ihre Bedeutung im Rahmen der Tumorabwehr in situ ist bislang noch unklar [1272].

Hemmung durch Antigen-Antikörperkomplexe

Es gibt Anhaltspunkte, daß Antigen-Antikörperimmunkomplexe den afferenten und efferenten Arm der Immunreaktionen gegen Tumoren deutlich inhibieren können. Diese Inhibition ist im afferenten

Schenkel der Immunreaktion verschiedenartig. Eine Spekulation (die auch eine Erklärung für die immunsuppressive Wirksamkeit von Antikörpern sein könnte) ist, daß spezifisch geprägte Lymphozyten mit ihren Antigenrezeptoren an freie Epitope des Antigens im Immunkomplex binden und zum einen dadurch «neutralisiert» werden; zum anderen werden die hieraus resultierenden Lymphozyten-Immunkomplexe über Fc-vermittelte Reaktionen der beteiligten Antikörper durch Makrophagen und Granulozyten phagozytiert und hierdurch die spezifisch geprägten Lymphozyten eliminiert [678, 679]. Es bestehen des weiteren Anhaltspunkte, daß Antigen-Antikörperkomplexe T-Suppressorzellen induzieren [460], möglicherweise durch Bindung an Fc-Rezeptoren [561, 999, 1134]. Geringere Konzentrationen der Immunkomplexe führen ähnlich wie freies Antigen zu einer spezifischen, hohe Immunkomplexkonzentrationen zu einer unspezifischen Blockade der T-Zellen über deren Fc-Rezeptoren [494].

Antigen-Antikörperkomplexe können durch Kreuzvernetzung von Antigen- und Fc-Rezeptoren B-Zellen blockieren (Übersicht bei [1313, 1314]). Hierdurch wird die Antikörperbildung von B-Zellen [1000, 1001] auch in der Abwesenheit von T-Zellen oder T-Zellfaktoren [1062] gehemmt. Bei T-Zell-abhängiger Antikörperantwort blockieren Immunkomplexe die T-Zell-B-Zellkooperation [641].

Der efferente Arm der Immunantwort wird durch Immunkomplexe, sowohl antigenspezifisch als auch unspezifisch, gehemmt. So zeigt das Serum von Tieren mit progressiv wachsenden Moloney-Sarkomavirus-induzierten Tumoren einen deutlichen Hemmeffekt auf die CTL. Im Gegensatz hierzu inhibiert Serum von Tieren, deren Tumor zur Regression gekommen ist, nicht und kann sogar den blockierenden Effekt des ersten Serums aufheben [576, 577]. Überführung der blockierenden Immunkomplexe in einen extremen Antigen- oder Antikörperüberschuß hebt die Hemmwirkung auf [50, 724].

Immunkomplexe können auch die LAI [540, 541] und die ADCC [914, 915] hemmen. Speziell bei der ADCC ist die Hemmung naturgemäß nicht antigenspezifisch. Im Serum von tumortragenden Tieren und von Tumorpatienten sind blockierende Immunkomplexe nachzuweisen [540, 580, 1388]. Ihre biologische Bedeutung ist letztendlich unklar. Sie mag jedoch durch die Beschleunigung des Wachstums eines Polyomavirus-induzierten Sarkoms der Ratte [65, 66] nach Injektion von Tumorantigen-Antikörper-Immunkomplexen veranschaulicht werden.

Ansatzpunkte zur Tumorimmuntherapie

Passive Immuntherapie mit Antikörpern

Die passive Immuntherapie geht von der Voraussetzung aus, daß der Tumorträger in seiner Immunreaktion insuffizient ist. Diese Insuffizienz zu beheben, wird angestrebt durch Gabe von spezifischen Antikörpern, durch Verabreichung von spezifisch geprägten Lymphozyten oder aber durch die Gabe von Lymphozytenmediatoren, von denen man erhofft, daß sie das zur effektiven Immunreaktion notwendige Wechselspiel anregen. Der Gabe von Antikörpern gegen TAA werden hierbei verschiedene therapeutische Wirkmechanismen zugrunde gelegt: zum einen eine komplementabhängige Zytolyse von Tumorzellen (siehe S. 34), die Initiierung einer Antikörper-vermittelten, zellulären Zytotoxizität (siehe S. 47) oder aber auch die Mithilfe bei der Abräumung von abgesonderten TAA und TAA-Immunkomplexen (siehe S. 34).

In der Tat kann bei einigen Virus-induzierten Tumoren wie beispielsweise dem Katzenlymphosarkom [298, 508], der Grossvirus-induzierten Leukämie bei AKR-Mäusen [1306], der Graffivirus-induzierten Leukämie [977], der Friend-Virus-Leukämie [240, 1145, 1265], dem Moloney-Virus-Sarkom [380] und bei Polyomavirus-induzierten Tumoren [66] durch die Verabreichung von Antikörpern gegen das jeweilige Virus-induzierte Antigen ein deutlicher Heilerfolg erzielt werden. Ähnliches konnte auch mit Antikörpern gegen chemisch induzierte Leukämien [977] oder gegen Thymus-Differenzierungsantigene [100] gefunden werden. Zytotoxisch wirksam waren sowohl syngene [380] als auch allogene [11, 380, 771, 858, 907, 1035, 1608] und xenogene [8, 277, 374, 409, 481, 940, 1075, 1086, 1394, 1610, 1619] Antikörper gegen Tumoren.

Besonders bei spontanen Tumoren wird die Möglichkeit der Therapie mit Antikörpern durch die Probleme eingeschränkt, Antikörper einer geeigneten Spezifität in genügend großer Menge herstellen zu können. Nur sehr hohen Dosen dieser Antikörper wird eine therapeutische Wirksamkeit nachgesagt [1006, 1152], möglicherweise wegen der bekannten unzulänglichen Konzentration dieser Antikörper im Tumorgewebe [642].

Mit der Technik, monospezifische «monoklonale» Antikörper in vitro herzustellen, besteht derzeit Hoffnung, das Spezifitäts- und Mengenproblem zu lösen. Monoklonale Antikörper mit Spezifität gegen TAA werden derzeit bereits eingesetzt für die Radioimmunszintigraphie, Immuntherapie und Radioimmuntherapie von Tumoren. Voraussetzung sowohl für die Immuntherapie als auch für die Radioimmuntherapie eines bestimmten Tumors ist die Fähigkeit eines aufgrund immunzytologischer und immunhistologischer Spezifitätsuntersuchungen ausgewählten monoklonalen Antikörpers (MAk), in vivo an einen Tumor mit ausreichender Menge zu binden [59, 75, 482]. Diese Bindung ist abhängig von der Menge des jeweiligen TAA, welche die in vivo wachsenden Tumorzellen an ihrer Membran exprimieren, von der Frage, wie variabel diese Expression auf den Tumorzellen ist, und ob und in welcher Menge dieses TAA bzw. das vom MAk auf dem TAA erkannte Epitop auch auf Normalgeweben vorkommt und für den MAk unter In-vivo-Bedingungen zugänglich ist. Des weiteren ist entscheidend, welche Affinität zum TAA und welche Penetrationsfähigkeit der jeweilige MAk besitzt. Da die Lokalisation sowohl tierexperimentell als auch klinisch mit radioszintigraphischen Methoden ermittelt wird, ist entscheidend, welches Isotop mit welcher die Antikörperfunktion (Antigenbindung) möglichst schonenden Methode an den Antikörper gebunden wird.

Ausgehend von den grundlegenden Arbeiten von Pressman und Sternberger [1168] werden für die Radioimmunszintigraphie heute meist die γ-strahlenden Radionuklide ^{131}J, ^{123}J oder ^{111}In nach direkter oder über Chelatbildner indirekter Kopplung an monoklonale Antikörper (Übersicht bei [1344]) verwendet. In Abhängigkeit von der Spezifität des MAk binden sich in tierexperimentellen Systemen (z. B. xenotransplantierte Humantumoren) ca. 2–10%, beim Menschen im Bereich von 0,1–1% des intravenös applizierten radioaktiv markierten Antikörpers über einen Zeitraum von 1–7 Tagen an den Tumor (Übersicht bei [75]). Der Zeitraum bis zur erfolgreichen Immunszintigraphie kann durch Verwendung von F(ab)$_2$-Fragmenten statt von ungespaltenem IgG drastisch auf einige Stunden verkürzt werden. Diese Verbesserung ist bedingt durch eine beschleunigte renale Ausscheidung nicht an den Tumor gebundener F(ab)$_2$-Fragmente und damit Verminderung der Hintergrundaktivität und durch eine höhere Gewebegängigkeit des F(ab)$_2$-Fragments [75; Bosslet et al. in Vorbereitung]. Eine wesentliche Erhöhung des am Tumor bindenden Anteils radioaktiv markierter MAk auf bis zu 40% ist auch klinisch durch eine regionale Applikation zu erzielen, beispielsweise in Form einer Injektion in die Leberarterie bei Lebermetastasen [354, 1012] oder einer intraperitonealen Applikation bei in der Bauchhöhle befindlichen Tumoren [354]. Zwar befindet sich die Radioimmunszintigraphie mit MAk klinisch noch im Versuchsstadium, doch weisen die bislang vorliegenden Ergebnisse bereits heute auf eine klinische Einsatzmöglichkeit bei der Tumorrezidivsuche in Ergänzung oder parallel zur Computertomographie hin [75]. Unklar ist jedoch immer noch, ab welcher Größe Tumoren mit der Radioimmunszintigraphie erkannt werden können, und Probleme geben weiterhin die begrenzte und variable Expression von TAA durch den Tumor, die Freisetzung von TAA in die Zirkulation mit Bildung von Immunkomplexen in der Peripherie durch den Testantikörper und der unterschiedliche Vaskularisationsgrad von Tumoren auf.

Bei Patienten, bei denen der Tumor radioimmunszintigraphisch erfaßt werden kann, besteht die Möglichkeit, durch Erhöhung der Dosis des radioaktiv markierten Antikörpers (von 1–5 mCi auf ca. 50–100 mCi) eine Radioimmuntherapie des Tumors zu versuchen. Im Gegensatz zur Immuntherapie mit «kalten» Antikörpern oder mit Immuntoxinen, d. h. mit Toxinen oder Zytostatika, die an MAk gekoppelt sind, haben radioaktiv markierte Antikörper den Vorteil, daß die Radiotoxizität nicht auf die das TAA exprimierenden Tumorzellen beschränkt ist, sondern auch benachbarte, gegebenenfalls TAA-negative Tumorzellen trifft. Die Selektion von therapieresistenten TAA-negativen Tumorzellen dürfte bei Anwendung der Radioimmuntherapie somit geringer sein (siehe Tabelle III).

Unter der Bedingung einer regionalen Applikation konnte die Radioimmuntherapie bereits erfolgreich bei Ovarialkarzinomen [354] (siehe Tabelle IV) und bei Lebermetastasen von Kolonkarzinomen [354] durchgeführt werden. Gute Aussichten bestehen, daß mit optimal «spezifischen» MAk gegen ausgewählte TAA und mit zunehmender Erfahrung auf dem Gebiet der regionalen Applikation (über die das jeweilige Organ versorgenden Arterien) Lungentumoren, Lebertumoren, Tumo-

Tabelle III. Tumortherapie mit Antikörpern

Präparation	Wirkmechanismen	Bemerkungen
Maus IgG (IgG_{2a}; IgG_3)	– Komplement-abhängige Zytotoxizität	
	– Antikörper-abhängige zelluläre Zytotoxizität	– Wirksamkeit abhängig von der Ausbildung tumor-assoziierter Antigene (TAA)
	– Induktion von autologen Anti-Idiotypen-Antikörpern	
Radioaktiv markierte Antikörper	– Radiotoxizität	– Radiotoxizität wirkt regional-lokal (nicht auf TAA-positive Tumorzellen beschränkt)
Immuntoxine, Immunzytostatika	– Zytotoxizität	– beschränkt auf TAA-positive Tumorzellen

Tabelle IV. Tumorimmuntherapie mit monoklonalen Antikörpern

	Antikörper	Klinische Wirksamkeit (Tumorregression)				
		100%	>50%	<50%	0	Progression
Houghton et al. [663] Melanom Stadium III/IV	Anti-GD_3 (IgG_3) intravenös appl.	–	3*	2	2	5
Epenetos et al. [355] Ovarialkarzinom Stadium III Bauchwassersucht (Aszites)	Anti-HMFG (IgG_1-^{131}J)	6 9	– –	– –	– 1	– –

* Anzahl der Patienten

ren der Gliedmaßen, der Niere und des Darmes der Radioimmuntherapie zugänglich werden.

Der Erfolg einer Immuntherapie von Tumoren mit «kalten» MAk

sollte entscheidend von der Fähigkeit des ausgewählten MAk abhängen, über die Aktivierung von Komplement und der ADCC zytostatisch für die das jeweilige TAA exprimierende Tumorzelle zu sein. Ein alternativer, heute in seiner Bedeutung noch nicht abzuschätzender Weg wäre die Hemmung von Rezeptoren für Wachstumshormone auf der Tumorzelle durch spezifische monoklonale Antikörper [927, 964]. MAk auch geeignet erscheinender Isotypen sind im allgemeinen relativ gering wirksam, sowohl bei der Komplement-abhängigen [585, 616, 682, 684] als auch bei der zellulär mediierten (ADCC) Zytolyse [585, 613, 682]. Die Kombination von mehreren monoklonalen Antikörpern scheint jedoch stärker zytolytisch zu sein [585]. Obwohl das Wachstum menschlicher Tumoren wie Melanome [823] oder Kolon-Karzinome [615] in Nacktmäusen durch MAk inhibiert werden konnte, bleiben Zweifel, ob Antikörper im allgemeinen und MAk im besonderen bei der Immuntherapie spontaner Tumoren einen Durchbruch bringen können. Auch hier wird die Art und die Anzahl von TAA-Molekülen auf der Tumorzelle und das Bindungsverhalten des MAk von Bedeutung sein. So wurde mit MAk gegen ein Protein-Antigen auf Melanomen bei Melanomen kein therapeutischer Einfluß auf das Tumorwachstum gesehen [1405]. Andererseits bewirkten MAk gegen ein Glykolipidantigen (GD_3) eine deutliche Regression von Melanomen [663] (siehe Tabelle IV).

In Fällen, wo therapeutische Wirksamkeit in experimentellen Systemen [100] oder bei klinischen Erkrankungen (vorwiegend bei Leukämien, Übersicht bei [873, 983]) nachgewiesen werden konnte, war hierzu das Fc-Teil des Antikörpers notwendig [1624] und die Subklasse IgG_{2a} der Klasse IgG_3 [100] und beide den Subklassen IgG_1 oder IgG_{2b} [615] oder IgM [873, 1354] überlegen.

In Entwicklung ist die Immunchemotherapie von Tumoren mit Hilfe von MAk, bei welcher die Antikörper als Träger mit immunologischer Spezifität für Zytostatika oder Toxine verwendet werden. Dieses Prinzip ist seit längerem bekannt [273, 337, 410, 465, 467, 931, 997] und hat durch die Entwicklung monospezifischer Antikörper einen neuen Auftrieb erfahren [121, 471, 836, 1526]. So konnten durch Kopplung von Ricin A an einen MAk gegen Tumor-assoziiertes Transferrin das Wachstum von Melanomen in Nacktmäusen inhibiert [1526] und experimentelle Leukämien therapiert werden [837].

Andere bereits experimentell untersuchte Toxine sind Gelonin, Diphterietoxin, Pokeweed Toxin und Purothionin (Übersicht bei [446]). Zusätzlich liegen bereits Erfahrungen mit Zytostatika wie Vinkaaloiden, Methotrexat und Anthrazyklinen, gekoppelt an MAk, vor (Übersicht bei [446, 1232]).

Jedoch auch dieser therapeutische Ansatz wird durch viele derzeit noch unwägbare Faktoren beeinträchtigt [61, 446, 642]. Hierzu gehören technische Schwierigkeiten bei der die Antikörperfunktion schonenden Kupplung von ausreichenden Mengen von Zytostatika an Antikörper, Probleme, genügend Zytostatika über die Antikörper an die Tumorzelle zu bringen, Probleme, bedingt durch unspezifische Lokalisation der Antikörper in vivo und durch die Abschilferung von Tumorantigenen von der Tumorzelle. Möglicherweise scheinen einarmige, Fc-haltige Antikörper (Fab-Fc-Fragmente) weniger die Abschilferung von Tumorantigenen, d. h. die Antigenmodulation auf der Zellmembran auszulösen und damit Vorteile in der Therapie aufzuweisen [476]. Trotz dieses Vorteils bleibt das Problem der vom Tumor gegebenenfalls im Überschuß in die Peripherie abgegebenen Tumorantigene bestehen [1019]. Trotz intensiver Bemühungen befinden sich Immuntoxine und Immunzytostatika immer noch in einem experimentellen Entwicklungsstadium. Es ist somit anzunehmen, daß die Zukunft von Immuntoxinen oder Immunzytostatika weniger in der direkten Patientenbehandlung als mehr in der Zerstörung von Tumorzellen bei der In-vitro-Behandlung von Knochenmark zur Autotransplantation in exzessiv zytostatisch behandelten Tumorpatienten liegen wird [623, 873].

Ein völlig neuer Aspekt in der Antikörpertherapie von Tumoren ist die aktive Bildung von Antikörpern gegen die Idiotypen des verabreichten Antitumorantikörpers [826, 1208]. Derartige Anti-Idiotypenantikörper sind bekanntermaßen spezifisch immunregulatorisch aktiv [342, 703, 704]. Nach Injektion von monoklonalen Antitumor-Antikörpern in Tumorpatienten war das Auftreten von Anti-Idiotypenantikörpern (deren variable Region eine Kopie der antigenen Determinante sein sollte, an welche der injizierte Antitumor-Antikörper bindet) überraschenderweise vergesellschaftet mit einer Regression der jeweiligen Tumoren [826]. Jedoch müssen weitere klinische Untersuchungen den hier zu vermutenden kausalen Zusammenhang belegen.

Passive Immuntherapie mit Lymphozyten und Produkten von Lymphozyten

Alternativ zu Antikörpern wurden Lymphozyten entweder von spezifisch immunisierten gesunden Spendern oder nach spezifischer In-vitro-Immunisierung zur Therapie von Tumoren verabreicht. So konnten chemisch induzierte Tumoren mit großen Mengen von derartig spezifischen, syngenen, allogenen oder xenogenen Lymphozyten therapiert werden (Übersicht [1066]). Jedoch auch das Gegenteil, nämlich eine Beschleunigung des Tumorwachstums, wurde sowohl in syngenen [677, 1162, 1536] als auch in allogenen Systemen [691] gefunden. Die unterschiedlichen Wirkungen waren abhängig von der Zeitdauer zwischen Sensibilisierung und Transfer der Lymphozyten, des weiteren von der Menge der transplantierten Lymphozyten. Geringe Mengen führten eher zu einer Beschleunigung, große Mengen zu einer Regression des Tumors [1162].

Die Schwierigkeiten, große Mengen syngener, spezifisch geprägter Lymphozyten für die Therapie von Tumorpatienten zur Verfügung zu stellen, sind der Grund gewesen, warum diese Therapieform klinisch nicht nennenswert geprüft wurde [1066]. Andererseits können allogene Lymphozyten wegen der Gefahr der Erzeugung einer Abstoßungsreaktion des Transplantats gegen den Wirt (Graft-versus-Host-Reaktion) nicht verwendet werden. Ein Ersatz für Lymphozyten scheint sich in der isolierten Ribonukleinsäure aus Lymphozyten von spezifisch mit dem jeweiligen Tumor immunisierten Tieren zu bieten. Diese sogenannte «Immun-RNA» soll nicht immunisierte Lymphozyten spezifisch für das jeweilige Antigen prägen und demzufolge sowohl in vitro als auch in vivo eine spezifische Immunreaktion gegen Antigene und gegen Tumoren induzieren [13, 439, 440, 761, 765, 766, 896, 1131, 1186, 1204]. So gibt es Anhaltspunkte für eine therapeutische Wirksamkeit von Immun-RNA bei experimentellen Tumoren [13, 441, 761, 896, 1574]. Es liegen jedoch auch negative Ergebnisse vor. Untersuchungen an chemisch induzierten Tumoren [441, 1273] zeigen, daß zu einer erfolgreichen Therapie neben der Immun-RNA Tumorantigene und Peritonealzellen bzw. Milzzellen (möglicherweise als APC) vonnöten sind, wobei besonders die Anzahl von Milzzellen für die Wirksamkeit entscheidend ist. Klinische Studien an unterschiedlichen Tumorpatienten sind derzeit im Gange, um die Verträglichkeit und Wirksamkeit von xenogener Immun-RNA bei Tumorpatienten zu belegen [425].

Neben Immun-RNA als antigenspezifischem Faktor wird der sogenannte Transfer-Faktor als mögliches Therapeutikum geprüft. Der Transfer-Faktor ist ein dialysierbarer Faktor aus Leukozyten [859], mit Hilfe dessen antigenspezifisch die Reaktion vom verzögerten Typ übertragen werden kann [1253]. Jedoch gibt es auch Anhaltspunkte für eine alternative oder zusätzliche unspezifische Stimulierung des Immunsystems [116, 327].

Die bislang vorliegenden Ergebnisse bei Tumorpatienten sind eher enttäuschend: In exakt kontrollierten Studien konnten bislang keine eindeutigen Veränderungen des Krankheitsverlaufs nachgewiesen werden [182, 1253], obwohl beispielsweise virale Erkrankungen bei Kindern mit Leukämie deutlich durch Transfer-Faktor beeinflußbar waren [1439].

Zu den Lymphozytenprodukten, welche besondere Bedeutung für Tumorimmuntherapie haben könnten, gehört das Interleukin II. Interleukin II ist ein Glykoprotein mit einem Molekulargewicht von 15 000 (Ratte, Mensch) bzw. 30 000 (Maus), welches von T-Helfer-Zellen sekretiert wird und nicht antigenspezifisch die Proliferation und Entwicklung von spezifisch geprägten Vorläufer-T-Zellen in Effektor-T-Zellen stimuliert [1240, 1406]. Zu diesen Effektor-T-Zellen gehören zytotoxische Lymphozyten. Eine Steigerung der Zytotoxizität von T-Zellen und von NK-Zellen gegen Tumorzellen wurde nach Gabe von Interleukin II bei Mäusen nachgewiesen [571, 1567]. Ähnliche Befunde konnten bereits auch schon bei Tumorpatienten nachgewiesen werden.

Ein weiterer nicht antigenspezifischer Mediator, dessen tumortherapeutische Wirkung bereits seit längerem präklinisch sowie klinisch untersucht wird, ist Interferon. Die zu Leukozyteninterferon oder Fibroblasteninterferon vorliegenden Daten (Übersichten bei [109, 783]) weisen auf verschiedene Wirksamkeitsmechanismen hin, so auf einen direkten inhibitorischen Effekt auf das Zellwachstum [1114, 1448], auf eine Beeinflussung der Zellmembran, die beispielsweise zu einer verstärkten Bindung von Antikörpern führt [881, 882] und auf eine immunmodulatorische Wirksamkeit. So sind die Antikörperantwort, die ADCC [609, 618], die CTL [880, 1645], die NK-Zytotoxizität [311, 312, 1347, 1519] und die Makrophagenphagozytose [666, 1294] meist erhöht, können jedoch auch erniedrigt sein, wahrscheinlich unter Bedingungen, bei denen die Inhibition der Zellteilung durch Interferon zum Tragen kommt. Zusätzlich kann eine Interferonbehandlung von Tumorzellen zu einer verstärkten Expression von TAA führen [517] und dadurch diese Tumorzellen für das Immunsystem angreifbarer machen. In experimentellen Tumoren wirken hohe Dosen von Interferon meist tumorstatisch, bewirken jedoch meist keine Tumorregression [109, 645]. Derartig wirksame hohe Dosierungen wurden bislang jedoch beim Menschen kaum angewandt. Klinische Ergebnisse liegen bereits vor, beispielsweise beim Brustkrebs, bei Lymphomen, Nierentumoren, Haarzelleukämien, Myelomen und Kehlkopfpapillomen (Übersicht bei [109; Wasielewski und Sedlack, Dt. med. Wschr, im Druck]). Zwar konnten bei einigen Tumorarten deutliche Regressionen gesehen werden (siehe Tabelle V), die ursprünglich beispielsweise mit dem α-Interferon gehegten Erwartungen, ein breit wirkendes Tumortherapeutikum in der Hand zu haben, wurden jedoch nicht erreicht. Dennoch besteht derzeit die Hoffnung, daß die experimentell unterschiedlich stark ausgeprägte antitumorale Wirksamkeit der verschiedenen Interferone (α-, β- und γ-Interferon) [247, 1254] auch klinisch ein unterschiedliches therapeutisches Ergebnis zeigt [109, 194].

Tabelle V. Möglichkeiten der Tumorimmuntherapie

Mediatoren	Klinische Untersuchungen
Interferon α (B-Ly; Null-Ly; Fibroblasten)	Nierentumoren ~30% Tumor- Haarzell-Leukämie ~50% regression chronische myeloische Leukämie (Erhaltungstherapie) Blasenkarzinom (lokal)
Interferon β (Epithelzellen; Fibroblasten)	im Gange (AIDS; Hautkrebs)
Interferon γ (T-Lymphozyten; NK-Zellen)	im Gange
Interleukin I (Makrophagen; (N)K-Zellen)	in Vorbereitung
Interleukin II (T-Lymphozyten; NK-Zellen)	im Gange (Kombination mit T-Zellen, aktivierten Killerzellen)
Tumornekrosisfaktor (T-Lymphozyten)	im Gange
Kolonie-stimulierende Faktoren	in Entwicklung (Verminderung der Knochenmarktoxizität von Zytostatika)
Galenische Zubereitung für die hochdosierte Langzeitapplikation	in Entwicklung
Prüfung von Kombinationen (IFα + IFγ; IL-2 + IFγ; TNF + IFγ)	in Vorbereitung

Aktive unspezifische Immunmodulation

Unter einer aktiven Immunmodulation wird eine Beeinflussung des Immunsystems eines Organismus über dessen unmodifizierte, funktionelle Kapazität verstanden. Sie ist als unspezifisch anzusehen, wenn ein mangelhaft oder normal arbeitendes Immunsystem nicht antigenspezifisch moduliert wird, und dieser Einfluß zu einer Unterdrückung, einer Normalisierung oder einer Verstärkung der primären und/oder sekundären Immunreaktion gegen jedwelches Antigen führt (Übersicht bei [1315, 1334]).

Bedingt durch die zum größten Teil nur marginalen Erfolge der konventionellen Therapie solider Tumoren und der offensichtlich werdenden Grenzen der Antibiotikatherapie hat die unspezifische Immuntherapie mit der Empfehlung von bakteriellen Ganzkeimpräparationen [544, 929, 932, 936, 1072] eine Wiederbelebung erfahren. Weitere Präparationen größtenteils bakterieller Herkunft kamen hinzu. In präklinischen Untersuchungen zeigen sie alle eine mehr oder weniger ausgeprägte immunmodulierende Aktivität auf die Zahl und/oder die Funktion von T-Zellen, B-Zellen, Makrophagen und/oder NK-Zellen (Übersichten bei [340, 846, 1194, 1224, 1315, 1334, 1364, 1569]).

Zugleich hemmt ein Großteil der Substanzen, prophylaktisch und in einigen Fällen auch therapeutisch gegeben, das Wachstum von Primärtransplantaten sowie von Metastasen experimenteller, sowohl allogener als auch syngener Tumoren (Übersicht bei [1352]). So wurde beispielsweise das Anwachsen i. v. injizierter maligner Zellen durch BCG [1549], Bordetella pertussis [14], Corynebacterium parvum [70, 832, 1170, 1602], Corynebacterium granulosum [975] und durch Ätherextrakte von Brucella abortus [1297] vermindert. Ähnliches konnte auch für spontane Metastasen nach Injektion von BCG [1649], C. parvum [832, 833, 1248, 1249] und Levamisole [1415] gefunden werden. Andererseits können BCG und hieraus hergestellter «Methanol extracted residue» das Wachstum von Tumoren beschleunigen [1622]. Ähnliches ist von CFA bekannt [32, 959]. Dennoch, die therapeutisch positiven Befunde führten zwangsläufig zur klinischen Prüfung der antitumoralen Wirksamkeit von Immunmodulatoren. Ersten eindrucksvollen klinischen Hinweisen [932, 933, 1560] folgten enttäuschende Ergebnisse (Übersichten bei [45, 462, 991, 1068, 1381, 1487, 1545]). Sie führten zu präparativen Arbeiten mit dem Ziel, die in den Ganzkeimpräparationen enthaltene immunpharmakologisch aktive Komponente zu isolieren, dadurch die Wirksamkeit der Präparation zu verstärken und ihre Nebenwirkungen zu vermindern.

Zahlreiche chemisch definierte Substanzen aus Pilzen oder Bakterien, des weiteren auch Antibiotika und verschiedene chemische Strukturen sind bislang in dieser Absicht isoliert und/oder synthetisiert worden (Übersicht bei [340, 1194, 1196, 1315, 1592]). Trotz ihrer Wirksamkeit in immunologischen Modellen zeigen auch diese Präparationen, soweit bislang nachweisbar, klinisch keine eindeutige tumortherapeutische Wirkung [622, 1381, 1487, 1545]. Eine Ausnahme besteht in der lokalen Applikation am oder im Tumor, die beispielsweise zur Regression von Melanomen oder zur Verhinderung von Lokalrezidiven beim Blasenkarzinom führt, jedoch ohne die systemische Tumorerkrankung wesentlich zu beeinflussen [193, 797, 854, 1214].

Dieser Widerspruch zwischen den präklinischen Daten und den bisherigen klinischen Ergebnissen hat zu einem Überdenken der Prädikativität der bislang benutzten Testmodelle zur Ermittlung von Immunmodulatoren geführt. Zusätzlich begründet er zwangsläufig Zweifel an der Hypothese, daß Tumorwachstum durch das Immunsystem kontrolliert wird und demzufolge über das Immunsystem beeinflußbar ist [1315]. Andererseits ist eine Reihe von Ursachen bereits bekannt, welche die antitumorale Wirksamkeit eines Immunsystems sowohl spezifisch als auch unspezifisch hemmen kann (siehe S. 49). Hierbei ist jedoch nicht auszuschließen, daß die verschiedenen Immunparameter, denen eine Rolle bei der Wachstumskontrolle von Tumoren nachgesagt wird, bei verschiedenen Tumoren nur Epiphänomene darstellen [1315]. Sollten sie jedoch wirklich einen entscheidenden Einfluß auf das Tumorwachstum haben, so bleibt die Frage offen, unter welchen Bedingungen und bei welchen Tumoren eine wirksame Immunantwort das Tumorwachstum hemmt oder fördert [168, 169, 702, 1162, 1165].

Aus dem Widerspruch der präklinischen zu den klinischen Daten haben sich Lehren sowohl für die Klinik [207] als auch für das «Screening» von neuen Substanzen ergeben. Diese Lehren werden in den derzeit bestehenden, experimentellen Systemen zum Auffinden neuer Immunmodulatoren in unterschiedlicher Weise berücksichtigt.

Das Screeningsystem des NCI (USA) beschränkt sich im wesentlichen auf diejenigen immunologischen Testsysteme, deren Meßparameter ein tumorzytolytischer oder tumorstatischer Effekt ist [1080, 1478]. Substanzen mit Wirksamkeit in diesen Testsystemen werden in verschiedenen Tumormodellen, bevorzugt auf die Wirksamkeit gegen spontane Metastasen, untersucht. Die Bewertung einer in diesen Modellen wirksamen Substanz erfolgt nach Abschätzung ihrer Toxizität.

Im Gegensatz hierzu berücksichtigen andere Screeningsysteme eine breitere Palette [830] oder (wie in unseren Laboratorien) alle [1315, 1333] derzeit wesentlichen Meßparameter für antigenspezifische und unspezifische Immunreaktionen. Immunmodulatorisch aktive Substanzen werden nachfolgend in den sogenannten Eignungsmodellen untersucht. Hierzu gehören Modelle für Autoimmunerkrankungen [302, 304, 1328], chronische Infektionen [302–306], Organtransplantations- und Entzündungsmodelle und verschiedene meist metastasierende Tumormodelle [1288, 1315].

In diesen Eignungsmodellen therapeutisch wirksame Substanzen werden auf toxikologische Nebenwirkungen, entsprechend den Empfehlungen der WHO [1622] für Adjuvanzien, geprüft und aus diesen Ergebnissen das Nutzen-Risiko-Verhältnis für den Patienten abgeschätzt. Am Beispiel des immunmodulatorisch aktiven Aminopeptidase-Inhibitors «Bestatin» bleibt abzuwarten, ob die klinische Prüfung das aus präklinischen Untersuchungen ersichtliche Wirkspektrum [138, 139, 301, 303, 306, 1284–1288, 1333] bestätigt und damit erstmals ein chemisch definierter, klinisch eindeutig wirksamer Immunmodulator zur Verfügung steht.

Beide erwähnten Screeningsysteme beinhalten die Wirksamkeitsprüfung in experimentellen Tumorsystemen. Die bisherige Erfahrung lehrt, daß die klinische Prädiktivität hier gewonnener Ergebnisse äußerst fragwürdig ist. Möglicherweise kann sie durch stärkere Berücksichtigung relevanter Parameter in den jeweiligen Modellen verbessert werden, beispielsweise durch eindeutige Verlängerung der Überlebenszeit oder aber auch durch Verwendung kliniknaher Tumormodelle, speziell auch streng syngener autochthoner Tumoren [398, 1334, 1478, 1479]. In diesem Zusammenhang können autochthone Tumoren an Haustieren und Nutztieren eine entscheidende Hilfe bieten [566, 1095, 1112, 1324, 1325].

Jedoch auch hier scheinen Ergebnisse ermittelt werden zu können, welche den bisherigen human-klinischen Erfahrungen bislang nicht entsprechen. So konnte nach i. v.-Verabreichung von BCG bei Hunden mit Osteosarkom eine Verzögerung [1094] sowie kein Effekt [1580, 1581] im Auftreten von Lungenmetastasen gesehen werden, während beim Mammakarzinom [146] ein therapeutischer Erfolg beobachtet wurde.

Andererseits führte die Injektion von BCG oder C. parvum intratumoral in Mammatumoren von Hunden nicht zu einer Verlängerung der Überlebenszeit [1112].

Aktive spezifische Immuntherapie

Aktive spezifische Tumorimmuntherapie beinhaltet die Behandlung eines Tumorpatienten mit einem Tumorantigen, von welchem vorausgesetzt wird, daß es immunologisch mit den Antigenen identisch oder kreuzreaktiv ist, welche sich auf den Tumorzellen des Patienten befinden. Eine weitere Voraussetzung ist, daß sich durch diese Impfung im Patienten eine spezifische Immunreaktion entwickelt, welche quantitativ und/oder qualitativ zum Ausgangszustand unterschiedlich ist, und welche das Wachstum und/oder die Neuentstehung von Tumorzellen inhibiert. Die Grundlagen für diese Art der Immuntherapie wurden in experimentellen Tumorsystemen und hier bevorzugt bei chemisch induzierten Tumoren erarbeitet. Ob die aktive spezifische Tumorimmuntherapie jedoch auch bei spontanen Tumoren, einschließlich der menschlichen Tumoren, therapeutisch wirkt, ist bislang noch nicht allgemeingültig belegt worden. Jedoch gibt es bei einigen Tumorerkrankungen und unter bestimmten Bedingungen bereits Hinweise für Wirksamkeit (Übersichten bei [1332, 1582]).

Abgeleitet von In-vitro-Erfahrungen bei der spezifischen immunologischen Zytolyse von Tumorzellen wurde postuliert, daß das Immunsystem in der Lage sei, ca. 10^5 Tumorzellen, was etwa 0,1 mg Tumorgewebe entsprechen mag, abzuwehren [930]. Diese Tumormasse ist klinisch bislang nicht diagnostizierbar [300], und ein derartiges Krankheitsstadium ist bei klinisch erfaßbaren Tumorerkrankungen nur durch Reduktion der Tumormasse durch palliative konservative Behandlungen, wie durch chirurgische, radiotherapeutische oder chemotherapeutische Methoden, zu erreichen. Spezifisch aktive Tumorimmuntherapie wird demzufolge vorzugsweise in Kombination mit derartigen bewährten Behandlungsmethoden anzuwenden sein.

Ein grundsätzliches Problem bei der spezifischen Tumorimmuntherapie ist die Frage nach dem geeigneten Tumorantigen bzw. der geeigneten Tumorantigenzubereitung. Speziell im Hinblick auf das zu erreichende Ziel sollte sie so immunogen wie möglich sein und gleichzeitig auch alle mit der Malignität assoziierten antigenen Determinanten tragen.

Für eine aktive Impfung zur Prophylaxe oder Therapie ist die Dosis des Tumorantigens entscheidend, speziell auch im Hinblick auf die Stimulation von T-Suppressorzellen nach Applikation kleiner Dosen [222, 817]. Hierbei kann sowohl die immunsuppressive als auch die protektive oder therapeutische Dosis von Tumor zu Tumor unterschiedlich sein [222]. Zusätzlich zur Dosis der Tumorantigenzubereitung sind des weiteren der Applikationsort und das Impfprogramm von entscheidender Bedeutung [334, 1156, 1622].

Inaktivierte Tumorzellen

Seit längerem ist bekannt, daß die Immunogenität eines Antigens mit seiner Größe korreliert [726] und des weiteren durch den (adjuvierenden) Trägeranteil entscheidend mitbestimmt wird [718, 787]. Weiterhin ist die Frage nach der chemischen Struktur des Tumorantigens bei einem Großteil der Tumoren ungeklärt. Aus diesen verschiedenen Gründen ist es demzufolge naheliegend gewesen, für die spezifische Tumorimmuntherapie ganze Tumorzellen zu verwenden.

Bekannt ist außerdem, daß lebende Tumorzellen stärker immunogen sind als inaktivierte Tumorzellen [1156], des weiteren, daß die durch Tumorzellen provozierte Immunantwort (gemessen an der Resistenz gegen das Wachstum transplantierter, teilungsfähiger Tumorzellen) von der Anzahl der verabreichten Zellen abhängig ist. In unterschiedlichen Tumorsystemen scheinen 10^7 bis 10^8 Tumorzellen, intradermal oder subkutan [694, 878, 879, 1154] in Abständen von 1 bis ca. 6 Wochen appliziert [26, 694, 892, 1014, 1154, 1155], die beste Wirkung zu erzielen. Andere Autoren empfehlen jedoch 10^4 Tumorzellen [1559]. Bei Applikation von syngenen oder autochthonen Tumorzellen ist eine Inaktivierung zur Vermeidung von Impftumoren notwendig. Zur Inaktivierung werden Bestrahlung oder In-vitro-Behandlung mit Zytostatika angewandt. Bestrahlungen sollen nicht zu stark sein, um die Membranantigene zu schonen [485, 694, 952]. So bestehen Anhaltspunkte, daß bereits mit einer Bestrahlungsdosis von 400 R Histokompatibilitätsantigene zerstört werden können [952]. Andererseits kann eine Bestrahlung von Tumorzellen deren vormals therapeutische Wirkung in eine das Tumorwachstum beschleunigende Wirkung verwandeln [1559]. Bei Zytostatikabehandlungen wird dem Mitomycin [89, 1206, 1626] der Vorzug gegeben. Hier scheint die Immunogenität der Tumorzellen optimal erhalten zu bleiben.

Die Injektion von inaktivierten Tumorzellen in Tumorträger kann deren Tumorwachstum nicht nur reduzieren [1156], sondern auch beschleunigen [353, 1087, 1088, 1170, 1559]. Ähnlich kann eine spezifische Immunisierung zu einer Wachstumsbeschleunigung nachfolgend transplantierter Tumorzellen führen [677, 692]. Nach Injektion eines Gemisches von Tumorzellen und spezifisch geprägten Lymphozyten vermehren sich die Tumorzellen dann besonders beschleunigt, wenn verhältnismäßig wenig Lymphozyten beigemischt worden sind [390, 956, 1162, 1164]. Dieses Phänomen führte zum Versuch, das Risiko einer Förderung des Tumorwachstums durch die aktive spezifische Tumorimmuntherapie durch Zumischung eines Adjuvans zu den Tumorzellen, wie beispielsweise C. parvum, zu vermindern [1170]. Andererseits gibt es Hinweise, daß die durch höhere Dosen eines Immunstimulators, beispielsweise von «Methanol-extracted-residue» von BCG verursachte Förderung des Tumorwachstums durch Kombination mit einer spezifischen Tumorimmuntherapie verhindert werden kann [1622].

So inhibiert ein Gemisch aus inaktivierten Tumorzellen und C. parvum in einigen experimentellen Tumorsystemen das Tumorwachstum [127, 1309] und die Entwicklung von Metastasen [1170]. Ähnliches konnte auch für Tumorzellgemische mit BCG gefunden werden [68, 244, 412, 547 – 550]. Entscheidend für die Wirksamkeit war die Menge an Tumorzellen, die Art und Menge des Adjuvans wie beispielsweise die Keimzahl des verwendeten BCG und Zeitpunkt und Intervall der Vakzinationen [549].

Diese Faktoren können von Tumor zu Tumor und in Abhängigkeit von der Konstitution und Kondition des Tumorträgers unterschiedlich sein, was ein Grund für die außerordentlich widersprüchlichen Ergebnisse der bislang vorliegenden klinischen Studien sein mag [253, 254, 725, 974, 1545]. Beispielsweise wurden beim Melanom (Stadium II) mit BCG in Kombination mit allogenen Tumorzellen Hinweise für eine therapeutische Wirkung gesehen [344], während eine gleichartige Studie, jedoch unter Verwendung autologer Tumorzellen, wegen einer starken, offensichtlich durch die Immuntherapie bedingten Förderung des Tumorwachstums abgebrochen werden mußte [950]. Beim Bronchialkarzinom war diese Form der Immuntherapie unter Verwendung von C. parvum unwirksam [1410], ebenso bei der akuten myeloischen Leukämie und unter Verwendung von BCG [44, 202, 448, 558, 886, 1008, 1150, 1595, 1596] oder C. parvum [356, 443], während beim Kolonkarzinom die Verabreichung inaktivierter autologer Tumorzellen im Gemisch mit BCG einen deutlichen therapeutischen Erfolg bewirkte [657].

Zellhomogenate

Zellhomogenate, hergestellt durch aufeinanderfolgendes Tiefgefrieren und Auftauen [232, 233, 473, 546, 878, 1153, 1308], durch Lyophilisieren [694, 1153, 1353] und durch Behandlung mit Ultraschall [402, 1014] oder Überdruck [1611] in der «French»-Presse [233, 1097], durch Stickstoff-Kavitation [1098] oder durch mechanische Zerkleinerung im Mörser [943, 1269], sind gleich [1353], meist jedoch geringer immunogen als intakte Zellen, wenn die Immunogenität durch Schutzversuche gegen eine nachfolgende Transplantation lebender Zellen bestimmt wurde [1156].

Alkoholextrakte von Tumorzellen [28, 29], Butanolextrakte [729] oder wasserlösliche Überstände nach mechanischer Tumorzellzerkleinerung [654, 728, 1005] erbrachten keine wesentlichen Vorteile gegenüber Ganzzellpräparationen. Extrakte von Tumorzellen mit hyperosmotischer (3 m-) KCl-Lösung [957] erwiesen sich als immunogen [958] und in vorläufigen Versuchen therapeutisch wirksam, wenn sie in einer Ölsuspension (Drakeol-Mineralöl) verabreicht wurden [198].

Tumorantigenpräparationen, gewonnen von Einzelzellen durch Extraktion mit stufenweise eingestellter isotonischer bis hypotonischer Kochsalzlösung [272] und nachfolgender Gelchromatographie, zeigten ebenfalls Immunogenität, ermittelt mit der DTH Reaktion [651]. Klinische Untersuchungen, vorwiegend an Lungentumorpatienten, haben bislang erste Hinweise für eine therapeutische Wirksamkeit ergeben [652, 653, 1449]. Ähnlich positive Ergebnisse wurden mit autologen, wasserlöslichen Extrakten von Tumorzellen erzielt, die mit Äthylchlorformiat polymerisiert worden waren und, vermischt mit PPD oder Candida albicans Antigen, als Adjuvans Patienten, speziell Hypernephrompatienten, verabreicht wurden [1476].

Ähnlich wie bei Verwendung von Tumorzellen konnte nach spezifischer Immunisierung mit unterschiedlichen Tumorzellpräparationen nicht nur ein Schutz gegen ein nachfolgendes Tumorzelltransplantat, sondern in manchen Fällen auch eine Beschleunigung des Transplantatwachstums beobachtet werden [208, 209, 1308, 1397, 1398]. Wurden Zellhomogenate mit Adjuvanzien vermischt, wie mit Freund's komplettem Adjuvans [402, 878] oder Bordetella pertussis [365, 366], konnte im allgemeinen eine Verbesserung der Immunogenität erreicht werden, jedoch auch das Gegenteil wurde beobachtet [878].

In klinischen Therapie-Untersuchungen mit Tumorzellen in CFA

[505], Zellhomogenaten alleine [674, 675] oder kombiniert mit CFA [404, 1448] wurden keine eindeutigen therapeutischen Ergebnisse gefunden. Eine neue Zubereitungsform scheint in einem Gemisch von Tumorzellhomogenaten, aufgenommen in einem Manganphosphatgel, zu bestehen, dessen Verträglichkeit bereits festgestellt wurde, dessen therapeutische Wirksamkeit aber erst noch ermittelt werden muß [1389].

Virusonkolysate

Onkolysate, erzeugt durch verschiedene Viren, zeigen gegenüber Zellhomogenaten aus nicht infizierten Tumorzellen eine deutlich stärkere Immunogenität. Derartig immunogene Onkolysate wurden mit West-Nil-Virus [822], unterschiedlichen Influenza-Virusstämmen [131–133, 420, 520, 531, 794, 887, 889], Newcastle disease virus [39, 103, 329], Vesicular stomatitis virus [531, 889], Friend-Leukämie-Virus [814] oder Vaccinia-Vakzine-Virus [1571] hergestellt. Poliovirus, Adenovirus und Herpesvirus erwiesen sich als unwirksam [39]. Voraussetzung für die Verwendung von Virusonkolysaten ist, daß die ausgewählten Viren für den tumortragenden Empfänger des Onkolysats nicht pathogen sind oder aber durch eine abschließende Behandlung des Onkolysats beispielsweise mit Formaldehyd [131] oder UV-Licht [420, 890] inaktiviert werden können.

Die Ursache der stärkeren Immunogenität von Virusonkolysaten scheint in den Virus-induzierten Antigenen zu liegen, die die Immunogenität von Membranantigenen der Tumorzelle im Sinne einer «helfenden Determinante» [133, 987] beeinflussen. Virusstrukturproteine als adjuvierende Komponenten konnten ausgeschlossen werden [520]. Die bislang vorliegenden klinischen Untersuchungen mit Virusonkolysaten bei verschiedenen Tumoren [420, 511, 1016, 1290, 1380, 1571, 1572] weisen zwar auf eine gute Verträglichkeit, jedoch nur bei einzelnen Patienten auch auf eine mögliche therapeutische Wirkung hin. Für eine endgültige Bewertung bleibt das Ergebnis von kontrollierten randomisierten prospektiven Studien abzuwarten [1571].

Chemisch behandelte Tumorzellen

In der Annahme, daß eine Tumorzelle, konjugiert an eine Fremdsubstanz, im Sinne eines Hapten-Carrierkomplexes besser immunogen wirken könne, wurden Proteine wie menschliches IgG [266], Kaninchen-IgG, Ziegen-IgG [942], Keyhole-limpet hemocyanin, Rinderserumalbumin oder methyliertes bovines IgG [157] an Tumorzellen gebunden und die Immunogenität derartiger Tumorzellen, teils vermischt mit CFA, durch Transplantationsexperimente bestimmt. In einigen Fällen, wie beim Mamma-Adenokarzinom [266] und beim MCA-induzierten Tumor der Maus [157], konnte ein erhöhter Schutz, bei anderen experimentellen Tumoren, wie beim Plasmazelltumor und Melanom der Maus und beim Mammakarzinom der Ratte, jedoch keine Wirksamkeitssteigerung nachgewiesen werden (Übersicht bei [1156]).

Klinische Fallstudien erbrachten keine eindeutigen oder bislang nicht reproduzierte Hinweise auf Wirksamkeit [266, 942].

Ein anderer Weg, die Immunogenität von Tumorzellen zu erhöhen, wurde durch Blockade von Schwefelwasserstoff-, Amino-, Karboxyl- und/oder Hydroxygruppen membranständiger Proteine und Glykoproteine versucht (Übersicht bei [1156, 1161]). Behandlungen von Tumorzellen wurden mit Jodazetat, Jodazetamid, N-Äthylmaleimid [26, 701, 866, 1156, 1573] oder Para-Hydroxymercuribenzoat [1156], des weiteren mit Dimethylsulfat oder Azetanhydrid [1433], Perjodat [1056, 1105], Concavalin A [353, 922, 1156], Dinitrophenyl [157, 922, 1153, 1612], DTIC [1040], Formaldehyd [111, 229, 473, 839, 878, 879, 1097, 1098] oder Glutaraldehyd alleine [88, 248, 430, 431, 743, 807, 1149, 1169, 1256, 1510] oder in Kombination mit Butylamin oder 1-Äthyl-3-(3-dimethylaminopropyl)-Karbodiimid [1433] durchgeführt.

Ein Vergleich der Wirksamkeit dieser Methoden ist nur eingeschränkt möglich, da unterschiedliche Tumormodelle benutzt wurden und die Ergebnisse zwischen starker Immunogenitätssteigerung und Wirkungslosigkeit schwanken. Dennoch herrscht der Eindruck vor, daß eine Steigerung der Immunogenität von Tumorzellen besonders durch Einführung neuer, als Haptene wirkender Gruppen wie beispielsweise durch p-Hydroxymercuribenzoat erreicht werden konnte [1156] oder aber durch eine Reduktion positiver (beispielsweise durch Acetanhydrid oder Aldehyde) oder negativer (z. B. durch Dimethylsulfat) Oberflächenladung [88, 1433]. Im Vergleich zu bestrahlten Zellen waren derartig chemisch modifizierte Zellen meist nicht tumorigen, aber deutlich stärker immunogen [1433].

Diese Immunogenitätssteigerung kam besonders zum Ausdruck, wenn wenig Zellen (zwischen 10^2 und 10^4) zur Prophylaxe verabreicht wurden, und war bei größerer Zellzahl (10^5 und 10^6) schwächer ausgeprägt [1433]. Die Immunogenitätssteigerung war des weiteren von der Anzahl der modifizierten Gruppen pro Zelle abhängig [1433]. Nicht nur die Immunogenität von Tumorzellen, sondern auch hiervon hergestellter Membranpräparationen konnte durch Behandlung mit den verschiedenen Reagenzien gesteigert werden [1434].

Die durch Belastungsversuche nachweisbare Immunogenitätssteigerung betraf im wesentlichen die zelluläre, weniger die humorale Immunreaktion [1153, 1156, 1433]. Mit Concavalin A behandelten (zusätzlich bestrahlten) Tumorzellen konnte nicht nur eine spezifische Prophylaxe, sondern auch eine wirksame spezifische Tumorimmuntherapie in verschiedenen experimentellen Tumorsystemen erzielt werden [353, 1206, 1627]. Die optimal wirksame Zelldosis lag im Bereich von 10^7 bis 2×10^7 Concavalin A behandelten Tumorzellen, mehrmals in Abständen von mehreren Tagen subkutan verabreicht [353]. Im Gegensatz hierzu führte die Verabreichung gleich vieler bestrahlter Zellen oder hiervon hergestellter entkernter Zellen, sogenannter «Zellgeister», mit oder ohne Jodazetamidbehandlung zu einer Beschleunigung des Tumorwachstums [353].

Eine weitere Möglichkeit, die Immunogenität von Tumorzellen zu erhöhen, wurde in der Veränderung der Fluidität der Membranlipide gesehen. Diesem Ansatz liegt der Gedanke zugrunde, daß Membranlipide die Expression von TAA auf Proteinen beeinflussen könnten. Als Hinweis hierfür mag der Befund gelten, daß eine Erhöhung der Viskosität der Membranlipide mit einer vermehrten Bindung von Antikörpern gegen membranständiges Rh(D)-Antigen einhergeht [1367], des weiteren daß Antigene, inkorporiert in Liposomen, stärker immunogen sind, wenn die Lipidschicht rigider ist [1634]. Dem Gedankengang zufolge wurde ein Cholesterolester (Cholesterylhemisuccinat) in die Membran unterschiedlicher Tumorzellen eingebaut und die Immunogenität bestimmt. Die Ergebnisse waren jedoch widersprüchlich. So wurde keine Wirksamkeit [458], aber auch eine starke Immunogenitätssteigerung beobachtet [1368]. Der Unterschied mag in dem von Zelltyp zu Zelltyp möglicherweise unterschiedlich schnellen Austritt der exogenen Lipide aus der Zellmembran in die jeweilige Umgebung begründet sein [458, 1365, 1367, 1413].

Mit Enzymen behandelte Tumorzellen

Tumorzellen wurden mit verschiedenen Enzymen behandelt, um ihre Immunogenität zu steigern. Mit Ausnahme von Neuraminidase

(Tab. VI) waren die Ergebnisse insgesamt jedoch enttäuschend. So konnte weder mit Kollagenase, Hyaluronidase, RNase oder DNase [76, 77, 114] noch mit Papain, Bromelain, Pancreatin, Ficin oder Chymotrypsin [114, 257, 258] eine Steigerung der Immunogenität von Tumorzellen (L1210-Leukämie oder SV40 Tumoren) erreicht werden. Bei Behandlung mit Trypsin waren die Ergebnisse widersprüchlich: Teils wurde eine Steigerung der Immunogenität erzielt [114], zum Teil zerstörte die Trypsinbehandlung eine bestehende Immunogenität von Tumorzellen, speziell auch dann, wenn sie bereits durch Vorbehandlung mit Neuraminidase verstärkt worden war [257]. Ähnlich konnte Killion [776, 777] durch Inkubation mit Glukosidase oder Protease die Immunogenität von Tumorzellen (auch nach deren VCN-Behandlung) zerstören, was auf Glykoproteine als immunogenes Agens auf der Zellmembran hinweist.

Tabelle VI. Spezifische Tumortherapie mit Neuraminidase

Klinische Versuche	Zahl der Studien	
	laufend	abgeschlossen
Leukämie (AML)		
(New York)		
Bekesi et al. [83]		2 (positiv)*
		1 (nicht signifikant)*
(Münster)		
Urbanitz et al. [1546]	1 (positiv)*	
Solide Tumoren		
(Schachbrettvakzination)		
Kolon		
(Wien)		
Wunderlich et al. [1628]		
Duke C		1 (positiv)*
Duke B	1	
Prostata		
(Gießen)		
Gutschank et al. [527]		
Stadium IV		1 (positiv)**
Stadium III	1	

* kontrollierte, randomisierte Studie
** Pilot-Studie

Arbeiten auf dem Gebiet der Zellmembranbiologie und Tumorimmunologie wurden hauptsächlich mit Neuraminidase von Vibrio cholerae (VCN) durchgeführt, weil dieses Enzym aufgrund seiner breiten Spezifität alle bekannten α-O-ketosidisch verbundenen Neuraminsäuren abspaltet (Übersichten bei [1320, 1332]), und des weiteren weil VCN ein induzierbares Exoenzym ist, was aus dem Kulturmedium von Vibrio cholerae relativ rein isoliert werden kann [995, 1270, 1271].

Grundlage der Tumorimmuntherapie mit Neuraminidase war die Beobachtung, daß verschiedene Zellen nach Behandlung mit VCN immunogener werden. So konnte in einer Reihe von Tumormodellen in der Maus und der Ratte gezeigt werden, daß im Vergleich zu unbehandelten Tumorzellen die Injektion von VCN-behandelten und in ihrer Proliferationsrate inhibierten Tumorzellen zu einer tumorspezifischen Widerstandsfähigkeit gegen das Wachstum nachfolgend transplantierter lebender Tumorzellen führt. Diese tumorspezifische Widerstandsfähigkeit war durch Transplantation von Lymphozyten immunisierter Tiere auf unbehandelte Tiere zu übertragen und demzufolge als Immunität zu bezeichnen (Übersichten bei [1320, 1332]).

Versuche, durch Applikation VCN-behandelter Tumorzellen nicht nur einen prophylaktischen Schutz, sondern auch einen therapeutischen Einfluß auf das Wachstum experimenteller Tumoren zu erzielen, erbrachten jedoch unterschiedliche Ergebnisse, nicht nur zwischen unterschiedlichen Untersuchergruppen, sondern auch bei ein und derselben Arbeitsgruppe und sogar in ein und demselben Tumormodell (Übersicht bei [1320, 1332, 1335]). Dennoch wurde auf der Basis positiver Befunde bei experimentellen Leukämien [76–78] die tumortherapeutische Wirkung VCN-behandelter Tumorzellen bei der akuten myeloischen Leukämie in Kombination mit der Chemotherapie geprüft und eine deutliche therapeutische Wirksamkeit in bezug auf Rezidivfreiheit und Überlebenszeitverlängerung ermittelt [81–83, 646, 648–650]. Die Befunde konnten durch eine andere Untersuchergruppe bestätigt werden [1546]. Anders war die Situation bei soliden Tumoren. Hier ergab sich in Kenntnis der Widersprüchlichkeit der Ergebnisse an experimentellen Tumoren die Frage, ob sich über die Erarbeitung des Wirksamkeitsmechanismus von VCN eine Erklärung für die Widersprüche im Therapieerfolg zum einen und Vorschläge für eine Optimierung der «Tumorimmuntherapie» zum anderen finden lassen.

Unstrittig ist bislang gewesen, daß Neuraminidase den Zell-zu-Zell-Kontakt durch Verminderung der negativen Oberflächenladung erhöht und hierdurch den immunologischen Erkennungsprozeß von Antigenen fördert (Übersichten bei [1320, 1586]). Fraglich war, ob die durch Neuraminidase bewirkte Demaskierung verborgener (Krypt-)Antigene (beispielsweise des Thomsen-Friedensreich-Antigens) wesentlich zum tumortherapeutischen Erfolg beiträgt [1323]. Tumortherapieerfolge mit VCN-behandelten Zellen sollten demzufolge besonders deutlich sein in solchen Spezies, bei denen eine Immunreaktion gegen Kryptantigene bereits vorliegt. Des weiteren war fraglich, ob durch Abspaltung von Neuraminsäure bereits exponierte tumorassoziierte Membranantigene zugänglicher und dadurch immunogener gemacht werden.

Am Modell der Rosettenbildung (zwischen Lymphozyten und Erythrozyten) nach VCN-Behandlung konnte bestätigt werden, daß VCN zu einer Verstärkung des Zell-zu-Zell-Kontaktes führt [1340]. Da dieser verstärkte Zell-zu-Zell-Kontakt durch Einführung von mehrfach, nicht jedoch von einfach sialylierten Gangliosiden in die Zellmembran der VCN-behandelten Lymphozyten aufzuheben war, kann gefolgert werden, daß die VCN-bedingte Verstärkung des Zell-zu-Zell-Kontaktes verursacht wird durch Verminderung der abstoßenden negativen Zellladung, verursacht durch Abspaltung von membranständiger N-Acetyl-Neuraminsäure. Zusätzlich war VCN jedoch auch in der Lage, Rezeptoren auf der Zellmembran, beispielsweise für das Fc-Fragment von IgG verstärkt freizusetzen [1318, 1340].

Im Gegensatz zu diesem die Annahme des verstärkten Zell-zu-Zell-Kontaktes bestätigenden Befunden konnten keine Anhaltspunkte gefunden werden, daß durch VCN-Behandlung Membranantigene (beispielsweise Membranimmunglobuline auf Lymphozyten [1318] und tumorassoziierte Antigene auf Bronchialkarzinomzellen [1336] oder Histokompatibilitätsantigene auf Trophoblasten [1376] zugänglicher beispielsweise für Antikörper werden. Ohne Zweifel demaskiert VCN durch Abspaltung von N-Acetyl-Neuraminsäure Kryptantigene wie das Thomsen-Friedenreich-Antigen — nicht nur auf Erythrozyten und verschiedenen Tumorzellen des Menschen, sondern auch auf Erythrozyten und Tumorzellen anderer Spezies [1336, 1533–1535]. Es liegen jedoch keine Hinweise vor, daß das T-Antigen von entscheidender Bedeutung bei der Tumortherapie mit VCN sein könnte. Eher das Gegenteil kann geschlußfolgert werden: So war die Tumortherapie mit VCN unwirksam in Transplantationstumoren, auf welchen relativ viel Thomsen-Friedenreich-Antigene nach VCN-Behandlung gefunden wurden, jedoch erstaunlicherweise wirksam beim Lewis-Lung-Adenokarzinom, welches, wenn überhaupt, nur geringe Mengen dieses Antigens aufweist [1336]. Des weiteren erwiesen sich beim Hund autologe Erythrozyten in Kombination mit VCN tumortherapeutisch als unwirksam [1329, 1336], obwohl auf ihnen, ähnlich wie bei den therapeutisch jedoch wirksamen Tumorzellen, nach VCN-Behandlung Thomsen-Friedenreich-Antigene nachzuweisen waren. Des weiteren ergaben sich keine Anhaltspunkte, daß der Erfolg der Tumortherapie mit Tumorzellen und VCN abhängt von der Menge an Thomsen-Friedenreich-Antigenen auf dem Mamma tumor des jeweiligen Hundes oder daß der Titer von Serumantikörpern gegen Thomsen-Friedenreich-Antigene durch diese Art der Therapie beeinflußt wird.

Mit unterschiedlichen Methoden konnte nachgewiesen werden, daß VCN an der Zellmembran haften bleibt [906, 955, 1047, 1316]. Diese Haftung schien aufgrund von unterschiedlichen Inhibitionsexperimenten zum einen nicht über den enzymatisch aktiven Teil des VCN-Moleküls zu gehen, zum anderen ergaben sich Anhaltspunkte, daß die Membranstruktur, an welche VCN bindet, ein Laktose- bzw. Galaktosehaltiges Glykoprotein ist [1335]. Da Galaktose den glykosidischen Bindungspartner von NeuAc darstellt und nach Abspaltung der NeuAc exprimiert wird, könnte die enzymatische Aktivität von VCN zu einer Vermehrung der Membranbindungsstellen für VCN führen. Der Befund, daß VCN an die Zellmembran bindet und gebunden dort enzymatisch aktiv ist, war aus zweierlei Gründen ausschlaggebend für die weiteren Versuche, den Wirkungsmechanismus von VCN in der Tumortherapie zu ermitteln. Zum einen stellt VCN ein exogenes, der Zellmembran aufsitzendes Antigen dar, welches die Antigenität der Zelle entscheidend verändern könnte [706, 710, 711]. Zum anderen besteht die Möglichkeit, daß membrangebundene VCN durch ihre enzymatische Aktivität eine adjuvierende Aktivität auf die Immunreaktion gegen das Antigen ausüben kann. Speziell die Adjuvansaktivität von VCN konnte sowohl für Zellen (Erythrozyten, Tumorzellen) als auch für Neuraminsäure-freie Antigene bewiesen werden [709–711, 811, 1319, 1322, 1343]. Sie war abhängig von der Menge des Antigens (Anzahl der Zellen), der enzymatischen Aktivität und der Menge an VCN, welche dem Antigen bzw. Immunogen zuzumischen war. So wirkte VCN gleichzeitig, aber getrennt vom Immunogen verabreicht, nicht steigernd auf die Immunantwort.

Als Zielzellen für die Adjuvansaktivität von VCN wurden Lymphozyten (T-Zellen) [709, 710, 711], nicht jedoch Makrophagen oder Granulozyten, identifiziert [1336]. Wahrscheinlich erhöht VCN die Reaktivität von Lymphozyten auf Immunmediatoren [809, 810]. In Tieren, deren Immunsystem bereits unspezifisch (d. h. nicht antigenspezifisch) stimuliert worden war, zeigte VCN keine Adjuvansaktivität [1332, 1335].

Aufgrund dieser Ergebnisse zum Wirkungsmechanismus von VCN wurde ein neues Applikationsschema für die Tumortherapie mit VCN entwickelt [1326, 1342]. Dieses auch angesichts des Befundes, daß in Experimenten mit autochthonen, in herkömmlicher Weise mit VCN behandelten Tumorzellen der Erfolg dieser Art der Therapie an der Ratte [1604] und am Mammatumor des Hundes [1317, 1327] von der applizierten Zelldosis deutlich abhängig war. Die therapeutisch wirksame Zelldosis lag hierbei nur fünffach unter derjenigen Zellzahl, welche eine Beschleunigung des Tumorwachstums bewirkte, und eine Erniedrigung der optimal wirksamen Zelldosis um den Faktor 10 hatte nur einen vorübergehenden therapeutischen Effekt.

Diese enge Abhängigkeit von der Zelldosis mag eine Erklärung für die widersprüchlichen Ergebnisse der Tumorimmuntherapie mit VCN-behandelten Tumorzellen sein. Andererseits ergab eine Abschätzung der an der Membran haftenden VCN-Menge, daß mit einer therapeutisch wirksamen Dosis VCN-behandelter Tumorzellen VCN in einer Größenordnung injiziert worden war, für welche im Gemisch mit einem Antigen Adjuvanswirksamkeit in den vorhergegangenen Experimenten nachgewiesen werden konnte. Zellen verschiedenen Ursprungs konnten auch unterschiedlich viel VCN an sich binden [906]. Es war somit zu vermuten, daß unterschiedliche Mengen an VCN (haftend an VCN-behandelten Tumorzellen) die Zelldosis-Abhängigkeit der therapeutischen Wirkung VCN-behandelter Tumorzellen in entscheidendem Maße mit verursacht haben könnte. Konsequenterweise wurde versucht, durch Applikation von Mischungen aus Tumorzellen und VCN eine standardisierbare Alternative zur Injektion von VCN-behandelten Tumorzellen zu etablieren. Hierzu wurden unterschiedliche Zahlen von autologen Tumorzellen mit jeweils unterschiedlichen Mengen an VCN vermischt und die jeweiligen Gemische gleichzeitig und getrennt voneinander Hunden mit Mammatumoren intradermal injiziert (Schachbrettvakzination; [1342]. Die Hoffnung war hierbei, daß die Stärke der DTH-Reaktion nicht nur für eine optimal antigene, sondern auch für eine immunogene Mischung als Meßparameter dienen könnte. Erstaunlicherweise ergab sich, daß diese Schachbrettvakzination selbst zumindest gleich tumortherapeutisch wirksam war wie die optimal therapeutisch wirksame Menge konventionell mit VCN behandelten Tumorzellen [1327]. Ähnliches konnte auch beim Lewis-Lung-Adenokarzinom der Maus gefunden werden: Auch hier erwies sich die Schachbrettvakzination therapeutisch als stärker wirksam als VCN-behandelte Tumorzellen oder als Einzelmischungen aus Tumorzellen und VCN [1328, 1336]. Als weiterer Vorteil der Schachbrettvakzination ergab sich, daß keinerlei Anzeichen einer Beschleunigung des Tumorwachstums gefunden wurden, obwohl je nach Ausbeute bei der Tumorzellpräparation Zellmengen injiziert worden waren, welche nach Behandlung mit VCN in konventioneller Weise und subkutan appliziert zu einer Verstärkung des Tumorwachstums geführt hatten. Insgesamt erwies sich die Schachbrettvakzination als verträglich; Nebenwirkungen konnten nicht beobachtet werden.

Die Gründe für die im Vergleich zu VCN-behandelten Tumorzellen verbesserte therapeutische Wirksamkeit der Schachbrettvakzination und für die hiermit verbundene Verringerung des Risikos einer behandlungsbedingten Beschleunigung des Tumorwachstums sind unklar. Obwohl die unterschiedlichen Adjuvansuntersuchungen es wahrscheinlich machen, daß die Schachbrettvakzination immunstimulierend wirkt, konnte diese Fragestellung durch tierexperimentelle Untersuchungen aus verschiedenen Gründen bislang nicht beantwortet werden. Erste klinische Ergebnisse am Kolonkarzinom zeigten anhand der Verstärkung der DTH-Reaktion im Verlauf wiederholter Schachbrettvakzination eine deutliche Immunstimulierung [1183, 1184]. Ob diese Immunstimulierung in kausalem Zusammenhang mit dem tumortherapeutischen Effekt steht, ist ungeklärt. Angesichts der derzeitigen Schwierigkeiten, Tumorantigene auf Tumorzellen zu identifizieren, und in Anbetracht der widersprüchlichen Ergebnisse über die In-situ-Bedeutung der verschiedenen Testmethoden zur Erfassung immunologischer Reaktionen gegen Tumorzellen bestehen Zweifel, ob ein kausaler Zusammenhang zwischen der Verabreichung eines spezifischen «Tumorimmuntherapeutikums», hiermit verbundener Veränderungen bekannter Immunparameter und einer Beeinflussung des Tumorwachstums mit den derzeit verfügbaren Methoden überhaupt erfaßt werden kann.

Nimmt man jedoch an, daß die Verabreichung von VCN-behandelten oder mit VCN vermischten Tumorzellen in einen Tumorträger zu einer spezifischen, das Tumorwachstum in situ hemmenden Immunreaktion führt, so gibt es keine eindeutige Erklärung für den beobachteten Vorteil der Schachbrettvakzination im Vergleich zu konventionell mit VCN behandelten Tumorzellen.

Vielleicht wirken zum einen bei der im Rahmen der Schachbrettvakzination erfolgten Kombination von möglicherweise stimulierend und supprimierend wirkenden Zelldosen die therapeutisch wirksamen Zelldosen dominant über andersartig wirkende Mischungen. Zum anderen könnte der individuellen Reaktionslage des Organismus durch die Schachbrettvakzination stärker Rechnung getragen werden als durch die Applikation einer einzelnen bestimmten Zellmenge.

Klinische Untersuchungen geben mittlerweile Anhaltspunkte, daß die Schachbrettvakzination bei ausgewählten Tumoren des Menschen eine relevante, therapeutische Wirkung hat [527, 1582, 1628]. Es bleibt abzuwarten, ob diese Wirkungen reproduzierbar sind und wir damit eine Möglichkeit der aktiven, antigenspezifischen Tumorimmuntherapie bei ausgewählten Tumoren in der Hand haben.

Zusammenfassung und Wertung

Die nun mehr als 80 Jahre andauernden Versuche, Tumorwachstum mit Hilfe von Maßnahmen, welche in das Immunsystem eingreifen, therapeutisch beeinflussen zu können, haben bislang nicht den gewünschten Erfolg gebracht. Wesentliche Probleme sind die Definition und Charakterisierung von Tumorantigenen und die Ermittlung derjenigen Immunreaktionen, welche in der Tumorabwehr entscheidend sind und (sollte die «Immune surveillance» zutreffen) vom «wachsenden» Tumorgewebe offensichtlich ausgeschaltet oder neutralisiert werden können.

Während bei chemisch induzierten tierexperimentellen und bei Virus-induzierten Tumoren tumorspezifische Antigene in der Tat nachgewiesen werden konnten, ist es bislang – auch unter Einbezug der Hybridomatechnologie und moderner biochemischer und immunologischer Analyseverfahren – noch nicht gelungen, tumorspezifische Antigene auf «spontanen» autochthonen Tumoren nachzuweisen. Vielmehr sind bislang auf der Zellmembran von Tumorzellen nur tumorassoziierte, embryonale organ- bzw. gewebespezifische oder kryptische Antigene mit vorwiegend serologischen Techniken identifiziert worden. Zusätzlich durchgeführte zellulärimmunologische Techniken, die im wesentlichen Zelltransformationen von Lymphozyten nach Kontakt mit Tumorzellen, zytotoxische Reaktionen von Lymphozyten gegen Tumorzellen und die (mit unterschiedlichen Methoden nachgewiesene) Mediatorausschüttung von Lymphozyten nach Kontakt mit Tumorzellen beinhalten, führten nicht zu einer weiteren Differenzierung von Tumorantigenen. Zudem entstanden Zweifel, ob mit Hilfe dieser Techniken (durchgeführt mit Serumantikörpern oder Lymphozyten vorwiegend isoliert aus dem Blut, seltener aus den jeweiligen Tumoren) bei Tumorpatienten die Erfassung einer antigenspezifischen, immunologischen Reaktionslage gegen ihre eigenen Tumoren möglich ist. Die Ergebnisse sind zu widersprüchlich; sie werden zusätzlich noch beeinträchtigt durch einen Mangel an (autologem) Tumormaterial zum einen und unspezifischen Kreuzreaktionen bei Verwendung von allogenem Tumormaterial zum anderen.

Auch die Entdeckung von nicht antigenspezifischen zytotoxischen Prinzipien, die vorwiegend gegen Tumorzellen und weniger gegen Normalzellen gerichtet sind, – wie die durch Makrophagen oder durch natürliche Killer-Zellen vermittelte Zytotoxizität – konnte nicht zur Klärung von Tumorantigenen beitragen. Ähnlich wie bei den humoralen und zellulären antigenspezifischen Reaktionsmechanismen ist die Rolle von Makrophagen und natürlichen Killer-Zellen bei der «Immune surveillance» von Tumoren zum Teil widersprüchlich und letztlich ungeklärt. Ähnliches gilt für die Antikörper-abhängige zelluläre Zytotoxizität.

Die verfügbaren immunologischen Methoden erlauben es jedoch, im Rahmen der gewachsenen Kenntnis von den Reaktionsmechanismen des Immunsystems und ihren Regelkreisen ausgewählte Aspekte der Wechselbeziehungen zwischen Tumoren und immunologischem Abwehrsystem speziell im Hinblick auf die Unterscheidung zwischen selbst und fremd zu studieren. Des weiteren konnte in diejenigen Mechanismen Einblick genommen werden, mit Hilfe derer Tumoren in der Lage zu sein scheinen, sich der Erkennung und dem Angriff des Immunsystems zu entziehen: so beispielsweise durch quantitative und qualitative Veränderung der Tumor(-assoziierten) Antigene, durch Neutralisation der spezifischen Abwehrmechanismen nach Abscheidung von Tumor(-assoziierten) Antigenen, durch Inhibition von Makrophagen und durch Aktivierung von Suppressorzellen. Ob diese, meist tierexperimentell und/oder durch In-vitro-Untersuchungen gewonnenen Mechanismen jedoch für das In-situ-Wachstum von spontanen, autochthonen Tumoren bedeutsam sind, ist ebenso letztlich unentschieden wie die Rolle der bislang definierten immunologischen zytotoxischen Prinzipien bei der Tumorabwehr. Die Phänomene, welche bislang beim Patienten als immunologische Reaktion gegen seinen Tumor interpretiert worden sind, könnten demnach durchaus Epiphänomene darstellen.

Für den Fall, daß grundsätzlich Wechselbeziehungen zwischen dem Tumor und dem Immunsystem im Sinne einer «Immune surveillance» existieren, ist unbekannt, ob immunologische Reaktionen, welche peripher, das heißt aus Blutproben (Serum; Blutzellen) oder an der Haut von Tumorpatienten mit unterschiedlichen Methoden erfaßt werden, wirklich Aussagen über die Situation am oder im Tumor erlauben. Des weiteren ist nicht auszuschließen, daß durchaus immunologische Reaktionsmechanismen gegen Tumoren bestehen, welche wir mit unseren bislang etablierten Methoden noch nicht erfassen konnten.

Tumortherapie mit der Zielsetzung einer Immuntherapie ist bislang zum größten Teil pragmatisch durchgeführt worden, zum einen durch Verabreichung von Tumorzellen oder Tumorzellaufbereitungen, mit der Absicht, eine möglicherweise bestehende Immunreaktion gegen den Tumor antigenspezifisch und aktiv zu verstärken, zum anderen durch eine allgemeine Aktivierung des Immunsystems mit Hilfe der nicht antigenspezifisch, d. h. unspezifisch wirkenden sogenannten Immunstimulanzien.

Eine weitere Methode ist die (passive) Substitution mit polyklonalen und in zunehmendem Maße mit monoklonalen Antikörpern gegen Tumorzellen, mit Immunmediatoren, beteiligt an der Regulation des Immunsystems (Interleukin II; Transfer-Faktor; Interferon) oder an der zytotoxischen oder zytostatischen Reaktion (Lymphotoxin, evtl. Interferon) oder mit «spezifisch» sensibilisierten Lymphozyten oder hieraus isoliertem genetischem Material (Immun-RNA) gewesen. Für alle diese Methoden und Präparate konnten tierexperimentelle Hinweise oder auch Beweise für eine tumortherapeutische Wirkung an ausgewählten Transplantationstumoren erbracht werden. Diese Wirkung war jedoch, soweit bereits geprüft, klinisch an autochthonen Tumoren nicht oder nur bei einer eingeschränkten Anzahl von Tumorarten oder bei nur wenigen Substanzen bzw. Behandlungsformen zu reproduzieren. Die Gründe hierfür sind mannigfaltig und letztlich in unserer mangelnden Kenntnis über die Wechselbeziehung zwischen Tumor und Immunsystem zu suchen. So sind die Fragen nach dem geforderten Wirkspektrum, nach der Dosis, dem Zeitintervall der Applikation, der Auswahl einer Begleittherapie und der Indikation und Kontraindikation in Abhängigkeit vom jeweiligen Reaktionszustand des Immunsystems und in bezug auf den Tumortyp rational nur dann für ein Tumorimmuntherapeutikum zu beantworten, wenn die für die angenommene «Immune surveillance» ausschlaggebenden immunologischen Mechanismen bekannt sind. Dieses gilt in gleicher Weise für Immunstimulanzien wie für die antigenspezifische aktive Tumorimmuntherapie. Jedoch besteht zwischen beiden Behandlungsprinzipien der Unterschied, daß bei der aktiven spezifischen Tumorimmuntherapie die Frage nach der Art des Immuntherapeutikums zumindest in allgemeiner Form beantwortet werden kann.

Nach heutiger Kenntnis sollte es eine Tumorantigen- bzw. Tumorzellaufbereitung sein, welche nach Verabreichung in den Tumorträgern deren für diesen Therapieansatz unterstellte, antigenspezifische (d. h. gegen den Tumor gerichtete) Immunreaktionen derartig verstärkt, daß trotz aller Variablen in der immunologischen Reaktionslage eine Regression des Tumorwachstums resultiert und das Risiko einer durch diese Behandlung bedingten Wachstumsbeschleunigung des Tumors auf ein Minimum reduziert wird. Da bei spontanen Tumoren das Tumorantigen bislang unbekannt ist, wird man bevorzugt die Tumorzellen selbst als Antigenträger verwenden. Tumorzellen haben zudem in tierexperimentellen Untersuchungen stärkere Immunogenität aufgewiesen als hieraus hergestellte Antigenaufbereitungen. Da widersprüchliche Angaben über die Bedeutung der bislang bekannten Immunmechanismen an der Insitu-Wachstumskontrolle von Tumoren bestehen, wird vorrangig zu prüfen sein, ob ein Verfahren zur spezifischen Tumorimmuntherapie zu einer Tumorregression führt. Zur Beantwortung dieser Fragestellung sind Transplantationstumoren aufgrund ihrer mangelnden Prädiktivität nur eingeschränkt geeignet. So waren Immunstimulanzien in diesen Modellen zwar deutlich therapeutisch wirksam, in zahlreichen klinischen Studien konnte diese tumortherapeutische Wirksamkeit jedoch nicht bestätigt werden. Entsprechend ausgedehnte klinische Untersuchungen, wie sie bereits mit Immunstimulanzien durchgeführt worden sind, fehlen bislang in der aktiven spezifischen Tumorimmuntherapie, bedingt durch die meist nur begrenzte Verfügbarkeit autologen oder entsprechend geeigneten Tumormaterials und aus Mangel an präklinischen Untersuchungen für verträgliche Tumorantigenaufbereitungen.

Aufgrund der bis heute vorliegenden, deutlich positiven klinischen Ergebnisse wäre es jedoch verfehlt, nicht weiter in die Entwicklung der aktiven antigenspezifischen Tumorimmuntherapie zu investieren. In ähnlicher Weise, wenn nicht noch betonter positiv, müssen die therapeutischen Möglichkeiten der passiven, spezifischen Tumortherapie mit monoklonalen Antikörpern, speziell auch in der Radioimmuntherapie, gesehen werden. Sowohl die präklinischen als auch die klinischen Ergebnisse befürworten ein weiteres starkes Engagement auf diesem Gebiet. Es ist ungewiß, ob mit Immunmediatoren, wie beispielsweise Interferon, Interleukin und TNF, größere Erfolge in der Tumortherapie erzielt werden können. Die anfänglichen Hoffnungen, mit Interferon ein breit wirkendes Tumortherapeutikum in der Hand zu haben, wurden enttäuscht. Trotz dieser Enttäuschung darf jedoch nicht übersehen werden, daß z. B. α-Interferon bei einigen wenigen Tumorarten eine deutliche therapeutische Wirkung hat. Ähnliches kann für die anderen Mediatoren zutreffen. Des weiteren ist die therapeutische Potenz von Kombinationen ausgewählter Immunmediatoren noch nicht erarbeitet worden. Hier bestehen begründete Hoffnungen, weitere umschriebene therapeutische Erfolge bei der einen oder der anderen Tumorerkrankung zu erzielen.

Insgesamt gesehen gibt es keine Anhaltspunkte, anzunehmen, daß die Immuntherapie die derzeitigen Therapieprobleme bei Tumoren in Form einer «Wunderdroge» lösen wird. Jedoch darf trotz aller bisherigen Enttäuschungen aufgrund der bislang vorliegenden Daten angenommen werden, daß bestimmte antigenspezifische und nicht-antigenspezifische Immuntherapeutika bei ausgewählten Tumoren einen therapeutischen Erfolg bringen werden, welcher mit anderen tumortherapeutischen Prinzipien nicht zu erreichen ist. Im ärztlichen Kampf mit der Tumorerkrankung würden hierbei auch kleine Erfolge zählen. Und kleine Erfolge wären der Beweis, daß man die richtige Richtung gewählt hat.

Danksagung

Ich danke den Herren Prof. W. Müller-Ruchholtz, Institut für Immunologie der Universität Kiel, Prof. V. Schirrmacher, Abteilung für zelluläre und Tumorimmunologie am Deutschen Krebsforschungszentrum in Heidelberg, und Prof. R. Kurth, Paul-Ehrlich-Institut in Frankfurt, für eine kritische Durchsicht der vorliegenden Arbeit.

Besonderer Dank gebührt Frl. Sylvia Freytag und Frl. Heike Waldschmidt, welche mit Umsicht, Geduld, außerordentlichem Können und Fleiß das Manuskript dieser Arbeit in eine leserliche Form verwandelt haben.

Literatur

1 Aalto, M.; Potila, M.; Kulonen, E.: The effect of silica-treated macrophages on the synthesis of collagen and other proteins in vitro. Expl Cell Res. *97:* 193 (1976).
2 Adams, D.O.; Snyderman, R.: Do macrophages destroy nascent tumors? J. natn. Cancer Inst. *62:* 1341 (1979).
3 Adams, D.O.; Johnson, W.J.; Marino, P.A.: Mechanisms of target recognition and destruction in macrophage-mediated tumor cytotoxicity. Fed. Proc. *41:* 2212 (1982).
4 Adams, D.O.; Hall, T.; Steplewski, Z.; Koprowski, H.: Tumors undergoing rejection induced by monoclonal antibodies of the IgG2a isotype contain increased numbers of macrophages activated for a distinctive form of antibody-dependent cytolysis. Proc. natn. Acad. Sci. USA *81:* 3506 (1984).
5 Adams, D.O.; Lewis, J.G.; Johnson, W.J.: Analysis of interactions between immuno-modulators and mononuclear phagocytes: Different modes of tumor cell injury require different forms of macrophage activation. Behring Inst. Mitt. *74:* 132 (1984).
6 Adorini, L.; Harvey, M.A.; Miller, A.; Sercarz, E.E.: Fine specificity of regulatory T cells. II. Suppressor and helper T cells are induced by different regions of hen egg-white lysozyme in a genetically nonresponder mouse strain. J. exp. Med. *150:* 293 (1979).
7 Aggarwal, B.B.; Kohr, W.J.; Hass, P.E.; Moffat, B.; Spencer, S.A.; Henzel, W.J.; Bringman, T.S.; Nedwin, G.E.; Goeddel, D.V.; Harkins R.H.: Human tumor necrosis factor. Production, purification, and characterization. J. biol. Chem. *260:* 2345 (1985).
8 Aizawa, M.; Imamura, T.; Motoyama, T.: On the immunological therapy of Yoshida sarcoma. Gann *42:* 138 (1951).
9 Alexander, P.: Foetal 'antigens' in cancer. Nature, Lond. *235:* 137 (1972).
10 Alexander, P.: Escape from immune destruction by the host through shedding of surface antigens: Is this a characteristic shared by malignant and embryonic cells? Cancer Res. *34:* 2077 (1974).
11 Alexander, P.; Connell, D.I.; Mikulska, Z.B.: Treatment of a murine leukemia with spleen cells or sera from allogeneic mice. Cancer Res. *26:* 1508 (1966).

12 Alexander, P.; Delorme, E. J.; Hall, J. G.: The effect of lymphoid cells from the lymph of specifically immunized sheep on the growth of primary sarcomata in rats. Lancet *i:* 1186 (1966).
13 Alexander, P.; Delorme, E. J.; Hamilton, L. D. G.; Hall, J. G.: Effect of nucleic acids from immune lymphocytes on rat sarcomata. Nature, Lond. *213:* 569 (1967).
14 Alexander, P.; Evans, R.; Grant, C. K.: Interplay of lymphoid cells and macrophages in tumor immunity. Annls Inst. Pasteur, Lille *122:* 645 (1972).
15 Alitalo, C.; Itovi, T.; Vaheri, A.: Fibronectin is produced by human macrophages. J. exp. Med. *151:* 602 (1980).
16 Allen, J. M.; Cook, G. M.: A study of the attachment phase of phagocytosis by murine macrophages. Expl Cell Res. *59:* 195 (1970).
17 Allison, A. C.; Harrington, J. S.; Birbeck, M.: An examination of the cytotoxic effects of silica on macrophages. J. exp. Med. *124:* 141 (1966).
18 Allison, A. C.; Law, L. W.: Effects of antilymphocyte serum on virus oncogenesis. Proc. Soc. exp. Biol. Med. *127:* 207 (1968).
19 Altevogt, P.; Fogel, M.; Cheingsong-Popov, R.; Dennis, J.; Robinson, P.; Schirrmacher, V.: Different patterns of lectin binding and cell surface sialylation detected on related high- and low-metastatic tumor lines. Cancer Res. *43:* 5138 (1983).
20 Amos, D. B.; Prioleau, W. H.; Hutchin, P.: Histochemical changes during growth of C3H ascites tumor PB8 in C57Bl mice. J. surg. Res. *8:* 122 (1968).
21 Anderson, A. C.: Parameters of mammary gland tumors in aging beagles. J. Am. vet. med. Ass. *147:* 1653 (1965).
22 Anderson, V.; Bjerrum, O.; Bendixen, G.; Schiodt, T.; Dissing, I.: Effect of autologous mammary tumor extracts on human leukocyte migration in vitro. Int. J. Cancer *5:* 357 (1970).
23 Anderson, R. J.; McBride, C. M.; Hersh, E. M.: Lymphocyte blastogenic responses to cultured allogeneic tumor cells in vitro. Cancer Res. *32:* 988 (1972).
24 Andrews, G. A.; Congdon, C. C.; Edwards, C. L.; Gengozian, N.; Nelson, B.; Vodopick, H.: Preliminary trials of clinical immunotherapy. Cancer Res. *27:* 2535 (1967).
25 Anglin, J. H., Jr.; Lerner, M. P.; Nordquist, R. E.: Blood group-like activity released by human mammary carcinoma cells in culture. Nature, Lond. *269:* 254 (1977).
26 Apffel, C. A.; Arnason, B. G.; Peters, J. H.: Induction of tumour immunity with tumour cells treated with iodoacetate. Nature, Lond. *209:* 694 (1966).
27 Appella, E.; Law, L. W.: Histocompatibility antigens and tumor-specific transplantation antigens. Expl Cell Biol. *44:* 131 (1976).
28 Aptekman, P. M.; Lewis, M. R.; King, H. D.: A method of producing in inbred albino rats a high percentage of immunity from tumors native in their strain. J. Immunol. *52:* 77 (1946).
29 Aptekman, P. M.; Lewis, M. R.; King, H. D.: Tumor-immunity induced in rats by subcutaneous injection of tumor extracts. J. Immunol. *63:* 435 (1949).
30 Arends, J. W.; Wiggers, T.; Schutte, B.; Thijs, C. T.; Verstijnen, C.; Hilgers, J.; Blijham, G. H.; Bosman, F. T.: Monoclonal antibody (1116 NS 19−9) defined monosialoganglioside (GICA) in colorectal carcinoma in relation to stage, histopathology and DNA flow cytometry. Int. J. Cancer *32:* 289 (1983).
31 Armstrong, M. Y. K.; Gleichmann, E.; Gleichmann, H.; Beldotti, L.; Andre-Schwartz, J.; Schwartz, R. S.: Chronic allogeneic disease. II. Development of lymphomas. J. exp. Med. *132:* 417 (1970).

Literatur

32 Asherson, G. L.; Allwood, G. G.: Depression of delayed hypersensitivity by pretreatment with Freund-type adjuvants. 1. Description of the phenomenon. Clin. exp. Immunol. *9:* 249 (1971).
33 Asherson, G. L.; Zembala, M.: T cell suppression of contact sensitivity in the mouse. III. The role of macrophages and the specific triggering of nonspecific suppression. Eur. J. Immunol. *4:* 804 (1974).
34 Atkin, N. B.: Premature chromosome condensation in carcinoma of the bladder: Presumptive evidence for fusion of normal and malignant cells. Cytogenet. Cell Genet. *23:* 217 (1979).
35 Ausiello, C.; Hokland, P.; Heron, I.: Interferon-induced augmentation of cytotoxic killer cell generation in mixed lymphocyte cultures: analysis of the effector cell product. Scand. J. Immunol. *13:* 263 (1981).
36 Ax, W.: Tumor diagnosis using electrophoretic mobility test (EMT): Review on state of the art with reference to the use of stabilized erythrocytes as indicator particles; in Flad, Herfarth, Betzler, Immunodiagnosis and immunotherapy of malignant tumor, p. 169, (Springer, Heidelberg 1979).
37 Ax, W.: Cell-mediated immunity model systems in tumor diagnosis. Behring Inst. Mitt. *65:* 94 (1980).
38 Ax, W.; Sedlacek, H. H.; Johannsen, R.: Antigenic specificities of human melanoma cells in vitro: Detection by xenogeneic antisera and HLA-isoantisera. Behring Inst. Mitt. *59:* 71 (1976).
39 Axler, D. A.; Girardi, A. J.: SV-40 tumor specific transplantation antigen in newcastle disease virus lysates of SV-40 transformed cells. Proc. Am. Ass. Cancer Res. *11:* 4 (1970).
40 Azuma, I.; Yamawaki, M.; Yasumoto, K.: Antitumor activity of Nocardia cell wall skeleton preparations in transplantable tumor in syngeneic mice and patients with malignant pleurisy. Cancer Immunol. Immunother. *4:* 95 (1978).
41 Babior, B. M.; Kipnes, R. S.; Curnatte, J. T.: Biological defense mechanisms: The production of leukocytes of superoxide, a potential bactericidal agent. J. clin. Invest. *52:* 741 (1973).
42 Bach, J. F.: Antigen-binding lymphocytes. Lancet *i:* 565 (1970).
43 Bach, J. F.; Dormont, J.; Dardenne, M.; Balner, H.: In vitro rosette-inhibition by antihuman antilymphocyte serum. Transplantation *8:* 265 (1969).
44 Baker, M. A.; Falk, J. A.; Taub, R. N.: Immunotherapy of human acute leukemia: antibody response to leukemia-associated antigens. Blood *52:* 469 (1978).
45 Baker, M. A.; Taub, R. N.; Carter, W. H.; Toronto Leukemia Study Group: Immunotherapy for remission maintenance in acute myeloblastic leukemia. Cancer Immunol. Immunother. *13:* 85 (1982).
46 Baldwin, R. W.: Immunity to methylcholanthrene-induced tumors in inbred rats following atrophy and regression of the implanted tumors. Br. J. Cancer *9:* 652: (1955).
47 Baldwin, R. W.: Tumour specific immunity against spontaneous rat tumours. Int. J. Cancer *1:* 257 (1966).
48 Baldwin, R. W.: Immunological aspects of chemical carcinogenesis. Adv. Cancer Res. *18:* 1 (1973).
49 Baldwin, R. W.: Tumour-specific antigens; in Homburger, The physiopathology of cancer, p. 334 (Karger, Basel 1974).

50 Baldwin, R. W.; Price, M. R.; Robins, R. A.: Blocking of lymphocyte-mediated cytotoxicity for rat hepatoma cells by tumour-specific antigen-antibody complexes. Nature 238: 185 (1972).
51 Baldwin, R. W.: Glaves, D.; Vose, B. M.: Embryonic antigen expression in chemically induced rat hepatomas and sarcomas. Int. J. Cancer 10: 233 (1972).
52 Baldwin, R. W.; Bowen, J. G.; Price, M. R.: Detection of circulation hepatoma D 23 antigen and immune complexes in tumour-bearer serum. Br. J. Cancer 28: 16 (1973).
53 Baldwin, R. W.; Embleton, M. J.; Price, M. R.: Inhibition of lymphocyte cytotoxicity for human colon carcinoma by treatment with solubilised tumour membrane fractions. Int. J. Cancer 12: 84 (1973).
54 Baldwin, R. W.; Price, M. R.; Robins, R. A.: Inhibition of hepatoma-immune lymph node cell cytotoxicity by tumour-bearer serum and solubilised hepatoma antigen. Int. J. Cancer 11: 527 (1973).
55 Baldwin, R. W.; Embleton, M. J.: Neo-antigens on spontaneous and carcinogen-induced rat tumors defined by in vitro lymphocytotoxicity assays. Int. J. Cancer 13: 433 (1974).
56 Baldwin, R. W.; Embleton, M. J.; Price, M. R.; Vose, B. M.: Embryonic antigen expression on experimental rat tumours. Transplant. Rev. 20: 77 (1974).
57 Baldwin, R. W.; Price, M. R.: Immunobiology of rat neoplasia. Ann. N. Y. Acad. Sci. 276: 3 (1976).
58 Baldwin, R. W.; Embleton, M. J.; Price, M. R.: Monoclonal antibodies specifying tumour-associated antigens and their potential for therapy. Mol. Aspects Med. 4: 329 (1981).
59 Baldwin, R. W.; Pimm, M. V.: Antitumor monoclonal antibodies for radioimmunodetection of tumors and drug targeting. Cancer Metastasis Rev. 2: 89 (1983).
60 Bale, W. F.; Spar, I. L.; Goodland, R. L.: Experimental radiation therapy with ^{131}I-carrying antibodies to fibrin. Cancer Res. 20: 1488 (1960).
61 Bale, W. F.; Contreras, M. A.; Grady, E. D.: Factors influencing localization of labeled antibodies in tumors. Cancer Res. 40: 2965 (1980).
62 Balfour, B. M.; Drexhage, H. A.; Kamperdijk, E. W. A.; Hoefsmit, E. C. M.: Antigen-presenting cells, including Langerhans cells, veiled cells and interdigitating cells. In: Microenvironments and cell differentiation. Ciba Foundation Symposium 84: 281 (1981).
63 Balkwill, F. R.; Taylor-Papadimitriou, J.: Interferon affects both G 1 and S + G 2 in cells stimulated from quiescence to growth. Nature, Lond. 274: 798 (1978).
64 Ballas, Z. K.; Feldbush, T. L.; Needleman, B. W.; Weiler, J. M.: Complement inhibits immune responses: C 3 preparations inhibit the generation of human cytotoxic T lymphocytes. Eur. J. Immunol. 13: 279 (1983).
65 Bansal, S. C.; Hargreaves, R.; Sjögren, H. O.: Facilitation of polyoma tumour growth in rats by blocking sera and tumour eluate. Int. J. Cancer 9: 97 (1972).
66 Bansal, S. C.; Sjögren, H. O.: Counteraction of the blocking of cell-mediated tumor immunity by inoculation of unblocking sera and splenectomy: Immunotherapeutic effects on primary polyoma tumors in rats. Int. J. Cancer 9: 490 (1972).
67 Barcinski, M. A.; Rosenthal, A. S.: Immune response gene control of determinant selection. I. Intramolecular mapping of the immunogenic sites on insulin recognized by guinea pig T and B cells. J. exp. Med. 145: 726 (1977).

Literatur

68 Bartlett, G. L.; Kreider, J. W.; Purnell, D. M.; Hockley, A. J., III: Treatment of visceral tumor with BCG-tumor cell vaccine. Cancer Immunol. Immunother. *4:* 15 (1978).
69 Basic, I.; Milas, L.; Grdina, D. J.; Withers, H. R.: Destruction of hamster ovarian cell cultures by peritoneal macrophages from mice treated with Corynebacterium granulosum. J. natn. Cancer Inst. *52:* 1839 (1974).
70 Basić, I.; Malenica, B.; Vujicić, N.; Milas, L.: Antitumor activity of Corynebacterium parvum administered into the pleural cavity of mice. Cancer Immunol. Immunother. *7.* 107 (1979).
71 Basombrio, M. A.: Search for common antigenicities among twenty-five sarcomas induced by methylcholanthrene. Cancer Res. *30:* 2458 (1970).
72 Bast, R. C., Jr.; Zbar, B.; Borsos, T. S.: BCG and cancer. New Engl. J. Med. *290:* 1413 (1974).
73 Bast, R. C., Jr.; Feeney, M.; Lazarus, H.; Nadler, L. M.; Colvin, R. B.; Knapp, R. C.: Reactivity of a monoclonal antibody with human ovarian carcinoma. J. clin. Invest. *68:* 1331 (1981).
74 Bean, M. A.; Pees, H.; Rosen, G.; Oettgen, H. F.: Prelabeling target cells with ^{3}H-proline as a method for studying lymphocyte cytotoxicity. Natn. Cancer Inst. Monogr. *37:* 41 (1973).
75 Begent, R. H. J.: Recent advances in tumour imaging: Use of radiolabelled antitumour antibodies. Biochim. biophys. Acta *780:* 151 (1985).
76 Bekesi, J. G.; St.-Arneault, G.; Holland, J. F.: Increase of leukemia L-1210 immunogenicity by Vibrio cholerae neuraminidase treatment. Cancer Res. *31:* 2130 (1971).
77 Bekesi, J. G.; St.-Arneault, G.; Walter, L.; Holland, J. F.: Immunogenicity of leukemia L 1210 cells after neuraminidase treatment. J. natn. Cancer Inst. *49:* 107 (1972).
78 Bekesi, J. G.; Roboz, J. P.; Walter, L.; Holland, J. F.: Stimulation of specific immunity against cancer by neuraminidasetreated tumor cells. Behring Inst. Mitt. *55:* 309 (1974).
79 Bekesi, J. G.; Holland, J. F.: Chemoimmunotherapy of leukemia in man and experimental animals. Int. Conf. on Immunotherapy of Cancer, Nov. 5–7, New York (1975).
80 Bekesi, J. G.; Roboz, J. P.; Holland, J. F.: Therapeutic effectiveness of neuraminidase-treated tumor cells as an immunogen in man and experimental animals with leukemia. Ann. N. Y. Acad. Sci. *277:* 313 (1976).
81 Bekesi, J. G.; Holland, J. F.: Active immunotherapy in leukaemia with neuraminidase-modified leukaemic cells. Recent results in cancer research, vol. 62, p. 78 (Springer, Berlin, Heidelberg 1977).
82 Bekesi, J. G.; Holland, J. F.: Immunotherapy of acute myelocytic leukaemia with neuraminidase-treated myeloblast and MER; in Rainer, Proc. Symp. «Immunotherapy of Malignant Diseases», p. 375 (Schattauer, Stuttgart, New York 1978).
83 Bekesi, J. G.; Holland, J. F.; Cuttner, J.; Silver, R.; Colman, M.: Immunotherapy of acute myelocytic leukemia with neuraminidase modified myeloblasts as immunogen. Proc. Am. Soc. Clin. Oncol. *1:* 36 (1982).
84 Benacerraf, B.: A hypothesis to relate the specificity of T lymphocytes and the activity of I region-specific IR genes in macrophages and B lymphocytes. J. Immunol. *120:* 1809 (1978).

85 Benacerraf, B.: Role of MHC gene products in immune regulation. Science *212:* 1229 (1981).
86 Benacerraf, B.; McDevitt, H. O.: Histocompatibility linked immune response genes. Science *175:* 273 (1972).
87 Benacerraf, B.; Germain, R. N.: Specific suppressor responses to antigen under I region control. Fed. Proc. *38:* 2053 (1979).
88 Ben-Efraim, S.; Ophir, R.; Relyveld, E. H.: Tumour immunoprophylaxis in mice using glutaraldehyde-treated syngeneic myeloma cells. Br. J. Cancer *43:* 554 (1981).
89 Benjamini, E.; Fong, S.; Erickson, C.; Leung, C. Y.; Rennick, D.; Scibienski, R. J.: Immunity to lymphoid tumors induced in syngeneic mice by immunization with mitomycin C-treated cells. J. Immunol. *118:* 685 (1977).
90 Bennink, J. R.: Negative selection studies of T help for cytotoxic T lymphocyte responses. Behring Inst. Mitt. *70:* 62 (1982).
91 Bentley, C.; Bitter-Suermann, D.; Hadding, U.; Brade, V.: In vitro synthesis of factor B of the alternative pathway of complement, activation by mouse peritoneal macrophages. Eur. J. Immunol. *6:* 393 (1976).
92 Berenson, J. R.; Einstein, A. B., Jr.; Fefer, A.: Syngeneic adoptive immunotherapy and chemoimmunotherapy of a friend leukemia requirement for thymus derived cells. J. Immunol. *115:* 234 (1975).
93 Bergholtz, B. O.; Thorsby, E.: HLA-D restriction of the macrophage-dependent response of immune human T lymphocytes to PPD in vitro: Inhibition by anti-HLA-DR antisera. Scand. J. Immunol. *8:* 63 (1978).
94 Berke, G.: Interaction of cytotoxic T lymphocytes and target cells; in Ishizaka, Prog. Allergy, vol. 27, p. 69 (Karger, Basel 1980).
95 Berke, G.: Cytotoxic T-lymphocytes. How do they function? Immunol. Rev. *72:* 5 (1983).
96 Berke, G.; Sullivan, K. A.; Amos, D. B.: Rejection of ascites tumor allograft. II. A pathway for cell-mediated tumor destruction in vitro by peritoneal exudate lymphoid cells. J. exp. Med. *136:* 1594 (1972).
97 Berke, G.; Amos, D. B.: Mechanisms of lymphocyte-mediated cytolysis. The LMC cycle and its role in transplantation immunity. Transplant. Rev. *17:* 71 (1973).
98 Bernards, R.; Schrier, P. I.; Houweling, A.; Bos, J. L.; Van der Eb, A. J.; Zijlstra, M.; Melief, C. J. M.: Tumorigenicity of cells transformed by adenovirus type 12 by evasion of T-cell immunity. Nature, Lond. *305:* 776 (1983).
99 Bernstein, I. D.; Zbar, B.; Rapp, H. J.: Impaired inflammatory responses in tumour bearing guinea pigs. J. natn. Cancer Inst. *49:* 1641 (1972).
100 Bernstein, I. D.; Nowinski, R. C.; Tam, M. R.; McMaster, B.; Houston, L. L.; Clark, E. A.: Monoclonal antibody therapy of mouse leukemia; in Kennett, McKearn, Bechtol, Monoclonal antibodies, p. 275 (Plenum Press, New York 1980).
101 Berzofsky, J. A.; Schechter, A. N.: The concepts of crossreactivity and specificity in immunology. Mol. Immunol. *18:* 751 (1981).
102 Bevan, M. J.: The major histocompatibility complex determines susceptibility to cytotoxic T cells directed against minor histocompatibility antigens. J. exp. Med. *142:* 1349 (1975).
103 Beverley, P. C. L.; Lowenthal, R. M.; Tyrrell, D. A. J.: Immune responses in mice to tumour challenge after immunization with newcastle disease virus-infected or X-irradiated tumour cells or cell fractions. Int. J. Cancer *11:* 212 (1973).

104 Beverley, P. C. L.; Lindh, D.; Delia, D.: The isolation of human haemopoietic progenitor cells using monoclonal antibodies. Nature 287: 332 (1980).
105 Bhakdi, S.; Tranum-Jensen, J.: Membrane damage by complement. Biochim. biophys. Acta 737: 343 (1983).
106 Bicker, U.: Immunomodulating effects of BM 12531 in animals and tolerance in man. Cancer Treat. Rep. 62: 1987 (1978).
107 Biddison, W. E.; Palmer, J. C.: Development of tumor cell resistance to syngeneic cell-mediated cytotoxicity during growth of ascitic mastocytoma P 815 Y. Proc. natn. Acad. Sci. USA 74: 329 (1977)
108 Biddle, C.: Stimulation of transplanted 3-methyl-cholanthrene-induced sarcomas in mice by specific immune and by normal serum. Int. J. Cancer 17: 755 (1976).
109 Billiau, A.: The clinical value of interferons as antitumor agents. Eur. J. Cancer clin. Oncol. 17: 949 (1981).
110 Bird, A. G. et al.: Cyclosporin A promotes spontaneous outgrowth in vitro of Epstein Barr virus-induced B-cell lines. Nature 289: 300 (1981).
111 Bismanis, J. E.: Immunization of mice against Ehrlich ascites carcinoma with formalinised tumour cells grown in tissue culture. J. Path. Bact. 87: 444 (1964).
112 Black, P. H.: The oncogenic DNA viruses: A review of in vitro transformation studies. Annu. Rev. Microbiol. 22: 391 (1968).
113 Black, P. H.; Rowe, W. P.; Turner, H. C.; Huebner, R. J.: A specific complement-fixing antigen present in SV 40 tumor and transformed cells. Proc. natn. Acad. Sci. USA 50: 1148 (1963).
114 Blakeslee, J. R., Jr.: The effect of various enzymes and chelating agents on the tumor specific transplantation antigen(s) (TSTA) of SV-40 induced tumor cells. Proc. Am. Ass. Cancer Res. 13: 31 (1972).
115 Bloch, K. J.: Heterogeneity in biologic functions of antibodies: Implications for immunologic tumor enhancement. Fed. Proc. 24: 1030 (1965).
116 Bloom, B. R.: Does transfer factor act specifically or as an immunologic adjuvant? New Engl. J. Med. 288: 908 (1973).
117 Bloom, B. R.; Bennett, B.; Oetttgen, H. F.; McLean, E. P.; Old, L. J.: Demonstration of delayed hypersensitivity to soluble antigens of chemically induced tumors by inhibition of macrophage migration. Proc. natn. Acad. Sci. USA 64: 1176 (1969).
118 Bloom, B. R.; Minato, N.; Neighbour, A.; Reid, L.; Marcus, D.: Interferon and NK cells in resistance to virus persistently infected cells and tumors; in Herberman, Natural cell-mediated immunity against tumors, p. 505 (Academic Press, New York 1980).
119 Blume, R. S.; Wolff, S. M.: The Chediak-Higasni syndrome: Studies in four patients and a review of the literature. Medicine, Baltimore 51: 247 (1972).
120 Blyden, G.; Handschumacher, R. E.: Purification and properties of human lymphocyte activating factor (LAF). J. Immunol. 118: 1631 (1977).
121 Blythman, H. E.; Casellas, P.; Gros, O.: Immunotoxins: Hybrid molecules of monoclonal antibodies and a toxin subunit specifically kill tumor cells. Nature 290: 145 (1980).
122 Boeri, D. G.: Europathologie: Recherches cliniques sur la respiration, sur le rire, sur le pleurer et sur le baillement des hémiplégiques. Gaz. heb. méd. Chir. 6: 73 (1901).

123 Boetcher, D. A.; Leonard, E. J.: Abnormal monocyte chemotactic response in cancer patients. J. natn. Cancer. Inst. *52:* 1091 (1974).
124 Bolhuis, R. L. H.; Schuit, H. R. E.; Nooyen, A. M.; Ronteltap, C. P. M.: Characterization of natural killer (NK) cells and killer (K) cells in human blood: Discrimination between NK and K cell activities. Eur. J. Immunol. *8:* 731 (1978).
125 Bolton, P. M.; Mander, A. M.; Davidson, J. M.; James, S. L.; Newcombe, R. G.; Hughes, L. E.: Cellular immunity in cancer: Comparison of delayed hypersensitivity skin tests in three common cancers. Br. med. J. *iii:* 18 (1975).
126 Bonney, R. J.; Naruns, P.; Davies, P.; Humes, J. L.: Antigen-antibody complexes stimulate the synthesis and release of prostaglandins by mouse peritoneal macrophages. Prostaglandins *18:* 605 (1979).
127 Bomford, R.: Active specific immunotherapy of mouse methylcholanthrene induced tumours with Corynebacterium parvum and irradiated tumour cells. Br. J. Cancer *32:* 551 (1975).
128 Boon, T.; Kellermann, O.: Rejection by syngeneic mice of cell variants obtained by mutagenesis of a malignant teratocarcinoma cell line. Proc. natn. Acad. Sci. USA *74:* 272 (1977).
129 Boon, T.; Van Pel, A.: Teratocarcinoma cell variants rejected by syngeneic mice: Protection of mice immunized with these variants against other variants and against the original malignant cell line. Proc. natn. Acad. Sci. USA *75:* 1519 (1978).
130 Boon, T.; Van Pel, A.; Uyttenhove, C.; Marchand, M.; Lambotte, P.: Tumor variants and the detection of tumor-specific transplantation antigens on spontaneous mouse tumors. Behring Inst. Mitt. *74:* 209 (1984).
131 Boone, C.; Blackman, K.; Brandchaft, P.: Tumor immunity induced in mice with cell-free homogenates of influenza virus infected tumor cells. Nature, Lond. *231:* 265 (1971).
132 Boone, C. W.; Blackman, K.: Augmented imunogenicity of tumor homogenates infected with influenza virus. Cancer Res. *32:* 1018 (1972).
133 Boone, C. W.; Paranjpe, M.; Orme, T.; Gillette, R.: Virus-augmented tumor transplantation antigens: Evidence for a helper antigen mechanism. Int. J. Cancer *13:* 543 (1974).
134 Borden, E. C.; Holland, J. F.; Dao, T.; Gutterman, J.; Wiener, L.; Chang, Y. C.; Patel, J.: Leukocyte-derived interferon in human breast carcinoma. The initial American Cancer Society phase II trial. Ann. intern. Med. *97:* 1 (1982).
135 Bosslet, K.; Schirrmacher, V.: Escape of metastasizing clonal cells or cell variants from tumor-specific cytolytic T lymphocytes. J. exp. Med. *154:* 557 (1981).
136 Bosslet, K.; Döring, N.; Sedlacek, H. H.: Effect of bestatin, an immunomodulator, on the generation of cytolytic T-lymphocytes (CTL) in vivo and in vitro and on natural killer (NK) cell activity in vivo. Immunobiol. *162:* 332 (1982).
137 Bosslet, K.; Müller, K.; Kurrle, R.; Sedlacek, H. H.: Murine monoclonal antibodies with specificity to tissue culture lines from human squamous carcinomas of the lung. Immunobiol. *162:* 332 (1982).
138 Bosslet, K.; Schirrmacher, V.; Sedlacek, H. H.: Comparison of the effects of bestatin and cyclosporin A on the cytolytic activity of anti-tumor and anti-H2 cytotoxic T lymphocytes in vivo and in vitro. Int. J. Immunopharm. *4:* 4 (1982).
139 Bosslet, K.; Schorlemmer, H. U.; Sedlacek, H. H.: Host cell mediated antimetasta-

tic effect of bestatin in a syngeneic T-lymphoma system. Proc. 13th Int. Cancer Congress, Washington (1982).
140 Bosslet, K.; Kern, H. F.; von Bülow, M.; Röher, H. D.; Klöppel, G.; Schorlemmer, H. U.; Kurrle, R.; Sedlacek, H. H.: A human monocyte cell surface antigen, highly expressed on an established pancreatic carcinoma cell line (TU II). Proc. Am. Pancreatic Association and The National Pancreatic Cancer Project, NCI, Chicago (1983).
141 Bosslet, K.; Kurrle, R.; Ax, W.; Sedlacek, H. H.: Monoclonal murine antibodies with specificity for tissue culture lines of human squamous carcinoma of the lung. Cancer Detect. Prevent. *6:* 181 (1983).
142 Bosslet, K.; Stark, M.; Kurrle, R.; Bischof, W.; Sedlacek, H. H.: Microenvironmental influences on the expression of monoclonal antibody defined immunochemically characterized membrane associated antigens on human small cell and adenocarcinomas of the lung. Proc. Int. Soc. for Oncodevelopmental Biol. Med., XI Annual Meeting, Stockholm (1983).
143 Bosslet, K.; Lüben, G.; Stark, M.; Sedlacek, H. H.: Molecular characteristics of two lung carcinoma cell-line associated membrane antigens. Behring Inst. Mitt. *74:* 27 (1984).
144 Bosslet, K.; Lüben, G.; Schwarz, A.; Hundt, E.; Harthus, H. P.; Seiler, F. R.; Muhrer, C.; Klöppel, G.; Kayser, K.; Sedlacek, H. H.: Immunohistochemical localization and molecular characteristics of three monoclonal antibody-defined epitopes detectable on carcinoembryonic antigen (CEA). Int. J. Cancer *36:* 75 (1985).
145 Bostock, D. E.: The prognosis following the surgical excision of canine mammary neoplasma. Eur. J. Cancer *11:* 389 (1975).
146 Bostock, D. E.; Gorman, N. T.: Intravenous BCG therapy of mammary carcinoma in bitches after surgical excision of the primary tumour. Eur. J. Cancer *14:* 879 (1978).
147 Boucheix, C.; Perrot, J.-Y.; Mirshahi, M.; Bernadou, A.; Rosenfeld, C.: A rapid method for detection of membrane antigens by immunofluorescence and its application to screening hybridoma antibodies. J. immunol. Methods *57:* 145 (1983).
148 Bowen, J. G.; Robins, R. A.; Baldwin, R. W.: Serum factors modifying cell-mediated immunity to rat hepatoma D 23 correlated with tumour growth. Int. J. Cancer *15:* 640 (1975).
149 Bowen, J. G.; Baldwin, R. W.: Tumour antigens and alloantigens. II. Lack of association of rat hepatoma D23-specific antigen with β 2-microglobulin. Int. J. Cancer *23:* 833 (1979).
150 Bowman, W. P.; Melvin, S. L.; Aur, R. J. A.; Mauer, A. M.: A clinical perspective on cell markers in acute lymphocytic leukemia. Cancer Res. *41:* 4794 (1981).
151 Boyse, E. A.; Old, L. J.; Luell, S.: Antigenic properties of experimental leukemias. II. Immunological studies in vivo with C 57 Bl/6 radiation-induced leukemias. J. natn. Cancer Inst. *31:* 987 (1963).
152 Boyse, E. A.; Stockert, E.; Old, L. J.: Modification of the antigenic structure of the cell membrane by thymus-leukemia (TL) antibody. Proc. natn. Acad. Sci. USA *58:* 954 (1967).
153 Braathen, L. R.; Thorsby, E.: Studies on human epidermal Langerhans cells. I. allo-activating and antigen-presenting capacity. Scand. J. Immunol. *11:* 401 (1980).

154 Braatz, J. A.; Scharfe, T. R.; Princler, G. L.; MacIntire, K. R.: Characterization of a human lung tumor-associated antigen and development of a radioimmunoassay. Cancer Res. *42:* 849 (1982).

155 Braatz, J. A.; Scharfe, T. R.; Princler, G. L.; McIntire, K. R.: Studies on a purified human lung tumor-associated antigen. Oncodev. Biol. Med. *3:* 169 (1982).

156 Braatz, J. A.; Hua, D. T.; Princler, G. L.: Serum levels of a human lung tumor-associated antigen using an improved radioimmunoassay. Cancer Res. *43:* 110 (1983).

157 Braun, W.; Plescia, O. J.; Raskova, J.; Webb, D.: Basic proteins and synthetic polynucleotides as modifiers of immunogenicity of syngeneic tumor cells. Israel J. med. Scis *7:* 72 (1971).

158 Braun, D. P.; Harris, J. E.: Modulation of the immune response by chemotherapy. Pharmacol. Ther. *14:* 89 (1981).

159 Braun, D. P.; Harris, J. E.: Serial immune function testing to predict clinical disease relapse in patients with solid tumors. Cancer Immunol. Immunother. *15:* 165 (1983).

160 Breard, J.; Fuks, A.; Friedman, S. M.; Schlossman, S. F.; Chess, L.: The role of P 23,30-bearing human macrophages in antigen-induced T lymphocyte responses. Cell. Immunol. *45:* 108 (1979).

161 Breard, J.; Reinherz, E. L.; O'Brien, C.; Schlossman, S. F.: Delineation of an effector population responsible for natural killing and antibody-dependent cellular cytotoxicity in man. Clin. Immunol. Immunopath. *18:* 145 (1981).

162 Brenner, B. G.; Jothy, S.; Shuster, J.; Fuks, A.: Monoclonal antibodies to human lung tumor antigens demonstrated by immunofluorescence and immunoprecipitation. Cancer Res. *42:* 3187 (1982).

163 Bretscher, M. S.; Raff, M. C.: Mammalian plasma membranes. Nature *258:* 43 (1975).

164 Britz, J. S.; Askenase, P. W.; Ptak, W.; Steinmann, R. M.; Gershon, R. K.: Specialized antigen presenting cells: Splenic dendritic cells and peritoneal exudate cells induced by mycobacteria activate T cells that are resistant to suppression. J. exp. Med. *155:* 1344 (1982).

165 Brodt, P.; Gordon, J.: Anti-tumor immunity in B lymphocyte-deprived mice. I. Immunity to a chemically induced tumor. J. Immunol. *121:* 359 (1978).

166 Broder, S.; Humphrey, R.; Durm, M. E.: Impaired synthesis of polyclonal (nonparaprotein) immunoglobulins by circulating lymphocytes from patients with multiple myeloma: Role of suppressor cells. New Engl. J. Med. *293:* 887 (1975).

167 Broder, S.; Muul, L.; Waldmann, T. A.: Suppressor cells in neoplastic disease. J. natn. Cancer Inst. *61:* 5 (1978).

168 Broder, S.; Waldmann, T. A.: The suppressor-cell network in cancer (first of two parts). New Engl. J. Med. *299:* 1281 (1978).

169 Broder, S.; Waldmann, T. A.: The suppressor-cell network in cancer (second of two parts). New Engl. J. Med. *299:* 1335 (1978).

170 Brown, J. P. et al.: Structural characterization of human melanoma-associated antigen p 97 using monoclonal antibodies. J. Immunol. *127:* 539 (1981).

171 Brooks, C. G.; Flannery, G. R.; Willmott, N.; Austin, E. B.; Kenwrick, S.; Baldwin, R. W.: Tumour cells in metastatic deposits with altered sensitivity to natural killer cells. Int. J. Cancer *28:* 191 (1981).

172 Brooks, C. G.; Urdal, D. L.; Henney, C. S.: Lymphokine-driven «differentiation» of cytotoxic T-cell clones into cells with NK-like specificity: Correlations with display of membrane macromolecules. Immunol. Rev. *72:* 43 (1983).
173 Brown, J. P.; Klitzman, J. M.; Hellström, K. E.: A microassay for antibody binding to tumor cell surface antigens using ^{125}I-labelled protein A from Staphylococcus aureus. J. Immunol. Methods *15:* 57 (1977).
174 Brown, J. P.; Nishiyama, K.; Hellström, I.; Hellström, K. E.: Structural characterization of human melanoma-associated antigen p97 with monoclonal antibodies. J. Immunol. *127:* 539 (1981).
175 Brown, J. P.; Woodbury, R. G.; Hart, C. E.; Hellström, I.; Hellström, K. E.: Quantitative analysis of melanoma-associated antigen p97 in normal and neoplastic tissues. Proc. natn. Acad. Sci. USA *78:* 539 (1981).
176 Brown, J. P.; Hewick, R. M.; Hellström, I.; Hellström, K. E.; Doolittle, R. F.; Dreyer, W. J.: Human melanoma-associated antigen p97 is structurally and functionally related to transferrin. Nature *296:* 171 (1982).
177 Brüggen, J.; Bröcker, E.-B.; Suter, L.; Redmann, K.; Sorg, C.: The expression of tumor-associated antigens in primary and metastatic human malignant melanoma. Behring Inst. Mitt. *74:* 19 (1984).
178 Brunda, M. J.; Herberman, R. B.; Holden, H. T.: Inhibition of murine natural killer cell activity by prostaglandins. J. Immunol. *124:* 2682 (1980).
179 Brunda, M. J.; Herberman, R. B. Holden, H. T.: Interferon-independent activation of murine natural killer cell activity; in Herberman, Natural cell-mediated immunity against tumors, p. 525 (Academic Press, New York 1980).
180 Brunda, M. J.; Herberman, R. B.; Holden, H. T.: Antibody-induced augmentation of murine natural killer cell activity. Int. J. Cancer *27:* 205 (1981).
181 Buinauskas, P.; McCredie, J. A.; Brown, E. R.; Cole, W. H.: Experimental treatment of tumors with antibodies. Archs Surg. *79:* 432 (1959).
182 Bukowski, R. M.; Deodhar, S.; Hewlett, J. S.; Greenstreet, R.: Randomized controlled trial of transfer factor in stage II malignant melanoma. Cancer *51:* 269 (1983).
183 Bullock, W. W.; Katz, D. H.; Benacerraf, B.: Induction of T-lymphocyte responses to a small molecular weight antigen. III. T - T cell interactions to determinants linked together: Suppression vs enhancement. J. exp. Med. *142:* 275 (1975).
184 Bullough, W. S.: Mitotic and functional homeostasis: A speculative review. Cancer Res. *25:* 1683 (1965).
185 Burakoff, S. J.; Finberg, R.; Glimcher, L.; Lemonnier, F.; Benacerraf, B.; Cantor, H.: The biologic significance of alloreactivity. The ontogeny of T-cell sets specific for alloantigens or modified self antigens. J. exp. Med. *148:* 1414 (1978).
186 Burch, P. R. J.: Carcinogenesis and cancer prevention. Nature, Lond. *197:* 1145 (1963).
187 Burdick, J. F.; Stemple, D.; Wells, S. A., Jr.: Study by the isotopic antiglobulin technique of a cross-reacting murine antibody produced by immunization with a syngeneic SV40 tumor. Int. J. Cancer *12:* 474 (1973).
188 Burdick, J. F.; Wells, S. A.; Herberman, R. B.: Immunologic evaluation of patients with cancer by delayed hypersensitivity reactions. Surgery Gynec. Obstet. *141:* 779 (1975).
189 Burger, M. M.: Surface changes in transformed cells detected by lectins. Fed. Proc. *32:* 91 (1973).

190 Burnet, M. F.: Cancer – A biological approach. IV. Practical applications. Br. med. J. *i:* 844 (1957).
191 Burnet, M. F.: Implications of cancer immunity. Aust. N. Z. J. Med. *1:* 71 (1973).
192 Busch, W.: Verhandlungen ärztlicher Gesellschaften. Berl. klin. Wschr. *5:* 137 (1868).
193 Camacho, F.; Pinsky, C.; Kerr, D.; Whitmore, W.; Oettgen, H.: Treatment of superficial bladder-cancer with intravesical BCG proceedings of the American Association of Cancer Research. Proc. Am. Ass. Cancer Res. *21:* 359 (1980).
194 Came, P. E.; Carter, W. A.: Interferon: Its application and future as antineoplastic agent. Drug Pharm. Sciences *24:* 301 (1984).
195 Cantanzaro, P. J.; Schwartz, H. J.; Graham, R. C., Jr.: Spectrum and possible mechanism of carrageenan cytotoxicity. Am. J. Path. *64:* 387 (1971).
196 Cantor, H.; Gershon, R. K.: Immunological circuits: Cellular composition. Fed. Proc. *38:* 2058 (1979).
197 Cantor, H.; Kasai, M.; Shen, F. W.; Leclerc, J. C.; Glimcher, L.: Immunogenetic analysis of «natural killer» activity in the mouse. Immunol. Rev. *44:* 3 (1979).
198 Cantrell, J. L.; McLaughlin, C. A.; Ribi, E.: Efficacy of tumor cell extracts in immunotherapy of murine EL-4 leukemia. Cancer Res. *39:* 1159 (1979).
199 Capel, P. J. A.; Preijers, F. W. M. B.; Allebes, W. A.; Haanen, C.: Treatment of chronic lymphocytic leukaemia with monoclonal antiidiotypic antibody. Neth. J. Med. *28:* 112 (1985).
200 Capron, A.; Dessaint, J.-P.; Haque, A.; Capron, M.: Antibody-dependent cell-mediated cytotoxicity against parasites: in Kallós, Prog. Allergy, vol. 31, p. 234 (Karger, Basel 1982).
201 Carbone, G.; Parmiani, G.: Immunogenic neoplastic clones derived in vitro from an originally non-immunogenic Balb/c fibrosarcoma. Tumori *63:* 513 (1977).
202 Carcassone, Y.; Favre, R.; Sebahoun, G.; Gartaut, J. A.; Imbert-Xeridot, L.: Chemo-immunotherapy done in maintenance treatment of acute non-lymphoblastic leukemia; in Mandelli, Therapy of acute leukemia (Lombardo editore, Roma 1977).
203 Carey, T. E.; Ko, L.; Takahashi, T.; Travassos, L. R.; Old, L. J.: AU cell surface antigen of human malignant melanoma. Solubilization and partial characterization. Proc. natn. Acad. Sci. USA *76:* 2898 (1979).
204 Carnaud, C.; Hoch, B.; Trainin, N.: Influence of immunologic competence of the host on metastasis induced by the 3LL Lewis tumor in mice. J. natn. Cancer Inst. *52:* 395 (1974).
205 Carrel, S.; Accolla, R. S.; Carmagnola, A. L.; Mach, J.-P.: Common human melanoma-associated antigen(s) detected by monoclonal antibodies. Cancer Res. *40:* 2523 (1980).
206 Carswell, E. A.; Old, L. J.; Kassel, R. L.; Green, S.; Fiore, N.; Williamson, B.: An endotoxin-induced serum factor that causes necrosis of tumors. Proc. natn. Acad. Sci. USA *72:* 3666 (1975).
207 Carter, S. K.: Biologic response-modifying agents: What is an appropriate phase I–II strategy? Cancer Immunol. Immunother. *8:* 207 (1980).
208 Casey, A. E.: A species limitation of an enhancing material derived from a mammalian tumor. Proc. Soc. exp. Biol. Med. *30:* 674 (1933).
209 Casey, A. E.: Proc. Soc. exp. Biol. Med. *31:* 663 (1934).

210 Castro, J. E.; Sadler, T. E., Jones, P. D.: Effects and mode of action of Corynebacterium parvum on murine tumor metastasis. Dev. Biol. Stand. *38:* 277 (1977).
211 Chang, R. S. M.: Continuous subcultivation of epithelial-like cells from normal human tissues. Proc. Soc. exp. Biol. Med. *87:* 440 (1957).
212 Chapman, H. A.; Hibbs, J. B.: Modulation of macrophage tumoricidal capability by components of normal serum: A central role for lipid. Science *197:* 282 (1977).
213 Chapman, H. A.; Hibbs, J. B.: Modulation of macrophage tumoricidal capability by polyene antibiotics: Support for membrane lipid as a regulatory determinant of macrophage function. Proc. natn. Acad. Sci. USA *75:* 4349 (1978).
214 Chauvenet, P. H.; Smith, R. T.: Relationship of tumor-specific transplantation antigens to the histocompatibility complex: Dissociation of in vitro alloantigen expression and in vivo alloimmunity from tumor-specific transplantation antigen strength. Int. J. Cancer *22:* 79 (1978).
215 Chee, D. P.; Townsend, C. M.; Galbraith, M. A.; Eilber, F. R.; Morton, D. L.: Selective reduction of human tumor cell populations by human granulocytes in vitro. Cancer Res. *38:* 4534 (1978).
216 Cheers, C.; Waller, R.: Activated macrophages in congenitally athymic 'nude' mice and lethally-irradiated mice. J. Immunol. *115:* 844 (1975).
217 Chervenick, P. A.; LeBuglio, A. F.: Human blood monocytes stimulators of granulocyte and mononuclear colony formation in vitro. Science *178:* 164 (1972).
218 Cheung, H. T.; Cantarow, W. D.; Sundharadas, G.: Characteristics of a low molecular weight factor extracted from mouse tumours that affects in vitro properties of macrophages. Int. J. Cancer *23:* 344 (1979).
219 Chia, D.; Terasaki, P. I.; Suyama, N.; Galton, J.; Hirota, M.; Katz, D.: Use of monoclonal antibodies to sialylated Lewis and sialylated Lewis for serological tests of cancer. Cancer Res. *45:* 435 (1985).
220 Chism, S. E.; Burton, R. C.; Warner, N. L.: Immunogenicity of oncofetal antigens: A review. Clin. Immunol. Immunopath. *11:* 346 (1978).
221 Churchill, W. H.; Piessens, W. F.; Sulis, C. A.; Davis, J. R.: Macrophages activated as suspension cultures with lymphocyte mediators devoid of antigen become cytotoxic for tumor cells. J. Immunol. *115:* 781 (1975).
222 Cihak, J.; Ziegler, H. W.; Kölsch, E.: Regulation of immune responses against the syngeneic ADJ-PC-5 plasmacytoma in Balb/c mice. II. Suppression of T-cell cytotoxicity by pretreatment of mice with subimmunogenic doses of tumour cells. Immunology *43:* 145 (1981).
223 Clark, R. A.; Klebanoff, S. J.: Neutrophil-mediated tumor cell cytotoxicity: role of the peroxidase system. J. exp. Med. *141:* 1442 (1975).
224 Clark, E. A.; Russell, P. H.; Egghart, M.; Horton, M. A.: Characteristics and genetic control of NK cell-mediated cytotoxicity activated by naturally acquired infection in the mouse. Int. J. Cancer *24:* 688 (1979).
225 Clark, I. A.; Virelizier, J.-L.; Carswell, E. A.; Wood, P. R.: Possible importance of macrophage-derived mediators in acute malaria. Infect. Immunity *32:* 1058 (1981).
226 Clarke, M. F., Gelmann, E. P., Reitz, M. S., Jr.: Homology of human T-cell leukaemia virus envelope gene with class I HLA gene. Nature, Lond. *305:* 60 (1983).
227 Cleveland, R. P.; Meltzer, M. S.; Zbar, B.: Tumor cytotoxicity in vitro by macrophages from mice infected with Mycobacterium bovis strain BCG. J. natn. Cancer Inst. *52:* 1887 (1974).

228 Coakham, H. B.; Kornblith, P. L.; Quindlen, E. A.: Autologous humoral response to human gliomas and analysis of certain cell surface antigens: In vitro study with the use of microcytotoxicity and immune adherence assays. J. natn. Cancer Inst. 64: 223 (1980).
229 Coca, A. F.; Dorrance, G. M.; Lebredo, M. G.: «Vaccination» in cancer. II. A report of the results of the vaccination therapy as applied in 79 cases of human cancer. Z. ImmunForsch. exp. Ther. 13: 543 (1912).
230 Cochran, A. J.; Grant, R. M.; Spilg, W. J.; Mackie, R. M.; Ross, C. E.; Hoyle, D. E.; Russel, J. M.: Sensitization to tumor-associated antigens in human breast carcinoma. Int. J. Cancer 14: 19 (1974).
231 Cocks, P.; Powles, R. L.; Chapuis, B.; Alexander, P.: Further evidence of response by leukaemia patients in remission to antigen(s) related to acute myelogenous leukaemia. Br. J. Cancer 35: 273 (1977).
232 Coggin, J. H.; Larson, V. M.; Hilleman, M. R.: Immunologic response in hamsters to homologous tumor antigens measured in vivo and in vitro. Proc. Soc. exp. Biol. Med. 124: 1295 (1967).
233 Coggin, J. H.; Larson, V. M.; Hilleman, M. R.: Prevention of SV40 virus tumorigenesis by irradiated, disrupted and iododeoxyuridine-treated tumor cell antigens. Proc. Soc. exp. Biol. Med. 124: 774 (1967).
234 Coggin, J. H.; Ambrose, K. R.; Anderson, N. G.: Fetal antigen capable of inducing transplantation immunity against SV40 hamster tumor cells. J. Immunol. 105: 524 (1970).
235 Coggin, J. H.; Ambrose, K. R.; Dierlam, P. J.: Proposed mechanisms by which autochthonous neoplasms escape immune rejection. Cancer Res. 34: 2092 (1974).
236 Colcher, D.; Horan Hand, P.; Nuti, M.: A spectrum of monoclonal antibodies reactive with mammary tumor cells. Proc. natn. Acad. Sci. USA 78: 3199 (1981).
237 Coley, W. B.: The treatment of malignant tumors by repeated inoculations of crysipelas, with a report of 10 original cases. Am. J. med. Sci. 105: 487 (1893).
238 Collavo, D.; Colombatti, A.; Chieco-Bianchi, L.; Davis, A. J. S.: Thymic lymphocyte requirement for murine sarcoma virus tumor prevention or regression. Nature, Lond. 249: 169 (1974).
239 Collavo, D.; Parenti, A.; Biasi, G.; Chieco-Bianchi, L.; Colombatti, A.: Secondary in vitro generation of cytolytic T-lymphocytes (CTL)s in the murine sarcoma virus system. Virus-specific CTL induction across the H-2 barriers. J. natn. Cancer Inst. 61: 885 (1978).
240 Collins, J. J.; Sanfilippo, F.; Tsong-Chou, L.; Ishizaki, R.; Metzgar, R. S.: Immunotherapy of murine leukemia. I. Protection against friend leukemia virus-induced disease by passive serum therapy. Int. J. Cancer 21: 51 (1978).
241 Collyn d'Hooghe, M.; Brouty-Boye, D.; Malaise, E. P.; Gresser, I.: Interferon and cell division. XII. Prolongation by interferon of the intermitolic time of mouse mammary tumour cells in vitro. Expl Cell Res. 105: 73 (1977).
242 Colten, H. R.: Biosynthesis of complement. Adv. Immunol. 22: 67 (1976).
243 Cone, R. E.: Molecular basis for T lymphocyte recognition of antigens; in Waksman, Prog. Allergy, vol. 29, p. 182 (Karger, Basel 1981).
244 Conti, G.; Zbar, B.; Asheley, M.; Rapp, H. J.; Hunter, J. T.; Ribi, E.: Eradication of microscopic metastases remaining after surgery by immunization with mycobacterial vaccines and tumor cells. Proc. Am. Ass. Cancer Res. 20: 220A (1979).

245 Cooper, H. S.: Peanut lectin-binding sites in large bowel carcinoma. Lab. Invest. *47:* 383 (1982).
246 Cove, D. H.; Woods, K. L.; Smith, S. C. H.; Burnett, D.; Leonard, J.; Grieve, R. J.; Howell, A.: Tumour markers in breast cancer. Br. J. Cancer *40:* 710 (1979).
247 Crane, J. L., Jr.; Glasgow, L. A.; Kern, E. R.; Youngner, J. S.: Inhibition of murine osteogenic sarcomas by treatment with type I or type II interferon. J. natn. Cancer Inst. *61:* 871 (1978).
248 Csaba, G.: Attempts to induce antitumour immunity with living attenuated cells. Neoplasma *14:* 167 (1967).
249 Cudkowicz, G.; Hochman, P. S.: Do natural killer cells engage in regulated reactions against self to ensure homeostasis? Immunol. Rev. *44:* 13 (1979).
250 Cullen, S. E.; David, C. S.; Shreffler, D. C.; Nathenson, S. G.: Membrane molecules determined by the H-2-associated immune response region: isolation and some properties. Proc. natn. Acad. Sci. USA *71:* 648 (1974).
251 Cullen, S. E.; Freed, J. H.; Nathenson, S. G.: Structural and serological properties of murine I a alloantigens. Transplant. Rev. *30:* 236 (1976).
252 Curries, G. A.: Masking of antigens on the Landschütz ascites tumor. Lancet *ii:* 1336 (1967).
253 Currie, G. A.: Eighty years of immnotherapy: A review of immunological methods used for the treatment of human cancer. Br. J. Cancer *26:* 141 (1972).
254 Currie, G. A.: Effect of active immunization with irradiated tumour cells on specific serum inhibitors of cell-mediated immunity in patients with disseminated cancer. Br. J. Cancer *28:* 25 (1973).
255 Currie, G. A.: Immunological aspects of host resistance to the development and growth of cancer. Biochim. biophys. Acta *458:* 135 (1976).
256 Currie, G. A.: Activated macrophages kill tumour cells by releasing arginase. Nature, Lond. *273:* 758 (1978).
257 Currie, G. A.; Bagshawe, K. D.: The effect of neuraminidase on the immunogenicity of the Landschütz ascites tumor. Site and mode of action. Br. J. Cancer *22:* 588 (1968).
258 Currie, G. A.; Bagshawe, K. D.: The role of sialic acid in antigenic expression: Further studies of the Landschütz ascites tumor. Br. J. Cancer *22:* 843 (1968).
259 Currie, G. A. Bagshawe, K. D.: tumor specific immunogenicity of methylcholanthrene-induced sarcoma cells after incubation in neuraminidase. Br. J. Cancer *23:* 141 (1969).
260 Currie, G. A.; Lejeune, F.; Fairley, G. H.: Immunization with irradiated tumour cells and specific lymphocyte cytotoxicity in malignant melanoma. Br. med. J. *ii:* 305 (1971).
261 Currie, G. A.; Basham, C.: Serum-mediated inhibition of the immunological reactions of the patient to his own tumor: A possible role for circulating antigen. Br. J. Cancer *26:* 427 (1972).
262 Currie, G. A.; Alexander, P.: Spontaneous shedding of TSTA by viable sarcoma cells: its possible role in facilitating metastatic spread. Br. J. Cancer *20:* 72 (1974).
263 Curry, R. A.; Quaranta, V.; Pellegrino, M. A.; Ferrone, S.: Serologically detectable human melanoma-associated antigens are not genetically linked to HLA-A and B antigens. J. Immunol. *122:* 2630 (1979).
264 Custer, R. P.; Eaton, G. J.; Prehn, R. T.: Does the absence of immunological sur-

veillance affect the tumor incidence in nude mice? First recorded spontaneous lymphoma in a nude mouse. J. natn. Cancer Inst. *51:* 707 (1973).

265 Cuttitta, F.; Rosen, S.; Gazdar, A. F.; Minna, J. D.: Monoclonal antibodies that demonstrate specificity for several types of human lung cancer. Proc. natn. Acad. Sci. USA *78:* 4591 (1981).

266 Czajkowski, N. P.; Rosenblatt, M.; Wolf, P. L.; Vasquez, J.: A new method of active immunization to autologous human tumour tissue. Lancet *ii:* 905 (1967).

267 Czarniecki, C. W.; Fennie, C. W.; Powers, D. B.; Estell, D. A.: Synergistic antiviral and antiproliferative activities of Escherichia coli-derived human alpha, beta, and gamma interferons. J. Virol. *49:* 490 (1984).

268 Damjanov, I.; Knowles, B. B.: Biology of disease: Monoclonal antibodies and tumor-associated antigens. Lab. Invest. *48:* 510 (1983).

269 Darzynkiewicz, Z.; Williamson, B.; Carswell, E. A.; Old, L. J.: Cell cycle-specific effects of tumor necrosis factor. Cancer Res. *44:* 83 (1984).

270 Dattwyler, R. J.: T cell antigens defined by monoclonal antibodies: A review. Plasma Ther. Transfus. Technol. *3:* 369 (1982).

271 Davey, G. C.; Currie, G. A.; Alexander, P.: Immunity as the predominant factor in determining metastases by murine lymphomas. Br. J. Cancer *40:* 590 (1979).

272 Davies, D. A.: Mouse histocompatibility isoantigens derived from normal and from tumor cells. Immunology *11:* 115 (1966).

273 Davies, D. A. L.; O'Neill, G. J.: In vivo and in vitro effects of tumor specific antibodies with chlorambucil. Br. J. Cancer *28:* suppl. 1, p. 285 (1973).

274 Davies, P.; Page, R. C.; Allison, A. C.: Changes in cellular enzyme levels and extracellular release of lysosomal acid hydrolases in macrophages exposed to group A streptococcal cell wall substance. J. exp. Med. *139:* 1262 (1974).

275 Davies, P.; Bonney, R. J.; Humes, J. L.; Kuehl, F. A., Jr.: The synthesis of arachidonic-acid oxygenation products by various mononuclear phagocyte populations; in van Furth, Mononuclear phagocytes, p. 1317 (Nijhoff, The Hague 1980).

276 Dawkins, H. J. S.; Shellam, G. R.: Augmentation of cell-mediated cytotoxicity to a rat lymphoma. I. Stimulation of non-T-cell cytotoxicity in vivo by tumour cells. Int. J. Cancer *24:* 235 (1979).

277 Day, E. D.; Pressman, D.: Purification of tumor-localizing antibodies. Ann. N. Y. Acad. Sci. *69:* 651 (1957).

278 Daynes, R. A.; Spellman, C. W.: Evidence for generation of suppressor cells by ultraviolet radiation. Cell. Immunol. *31:* 182 (1977).

279 Dean, J. H.; McCoy, J. L.; Cannon, G. B.; Leonard, C. M.; Perlin, E.; Kreutner, A.; Oldham, R. K.; Herberman, R. B.: Cell-mediated immune responses of breast cancer patients to autologous tumor-associated antigens. J. natn. Cancer Inst. *58:* 549 (1977).

280 De Baetselier, P.; Katzav, S.; Gorelik, E.; Feldman, M.; Segal, S.: Differential expression of H-2 gene products in tumor cells is associated with their metastatogenic properties. Nature, Lond. *288:* 179 (1980).

281 De Baetselier, P.; Gorelik, E.; Eshnar, Z.; Ron, Y.; Katzav, S.; Feldman, M.; Segal, S.: Metastatic properties conferred on non-metastatic tumors by hybridization of spleen B lymphocytes with plasmacytoma cells. J. natn. Cancer Inst. *67:* 1079 (1981).

Literatur

282 Deems, R. A.; Eaton, B. R.; Dennis, E. A.: Kinetic analysis of phospholipase A_2 activity toward mixed micelles and its implications for the study of lipolytic enzymes. J. biol. Chem. *250:* 9013 (1975).
283 Defendi, V.: Effect of SV40 virus immunization on growth of transplantable SV40 and polyoma virus tumors in hamsters (meeting abstract). Proc. Soc. exp. Biol. Med. *113:* 12 (1963).
284 Defendi, V.: Transformation in vitro of mammalian cells by polyoma and simian 40 viruses; in Homburger, Prog. exp. Tumor Res., vol. 8, p. 125 (Karger, Basel 1966).
285 Defendi, V.; Gasic, G.: Surface mucopolysaccharides of polyoma virus transformed cells. J. cell. comp. Physiol. *62:* 23 (1963).
286 Deichman, G. I.: Immunological aspects of carcinogenesis by deoxyribonucleic acid tumor viruses. Adv. Cancer Res. *12:* 101 (1969).
287 Deichman, G. I.; Kluchareva, T. E.: Loss of transplantation antigen in primary simian virus 40-induced tumors and their metastases. J. natn. Cancer Inst. *36:* 647 (1966).
288 De Landazuri, M. O.; Kedar, E.; Fahey, J. L.: Antibody-dependent cellular cytotoxicity to a syngeneic Gross virus-induced lymphoma. J. natn. Cancer Inst. *52:* 147 (1974).
289 De Lustro, F.; Haskill, J. S.: In situ cytotoxic T cells in a methylcholanthrene-induced tumor. J. Immunol. *121:* 1007 (1978).
290 DeMaeyer, E.; Mobraaten, L.; DeMaeyer-Guignard, J.: Prolongation par l'interferon de la survie des greffes de peau chez la souris. C. r. hebd. Séanc. Acad. Sci., Paris *D227:* 2101 (1973).
291 DeMaeyer, E.; DeMaeyer-Guignard, J.: Host genotype influences immunomodulation by interferon. Nature *284:* 173 (1980).
292 Deng, C.; El-Awar, N.; Cicciarelli, J.: Cytotoxic monoclonal antibody to a human leiomyosarcoma. Lancet *i:* 403 (1981).
293 Dennert, G.: Cloned lines of natural killer cells. Nature *287:* 47 (1980).
294 Dennis, J.; Donaghue, T.; Florian, M.; Kerbel, R. S.: Apparent reversion of stable in vitro genetic markers detected in tumour cells from spontaneous metastases. Nature, Lond. *292:* 242 (1981).
295 Dennis, J. W.; Donaghue, T. P.; Kerbel, R. S.: An examination of tumor antigen loss in spontaneous metastases. Invasion Metastasis *1:* 111 (1981).
296 Dennis, J. W.; Kerbel, R. S.: Characterization of a deficiency in fucose metabolism in lectin-resistant variants of a murine tumor showing altered tumorigenic and metastatic capacities in vivo. Cancer Res. *41:* 98 (1981).
297 Denis, J.; Waller, C.; Timpl, R.; Schirrmacher, V.: Surface sialic acid reduces attachment of metastatic tumour cells to collagen type IV and fibronectin. Nature, Lond. *300:* 274 (1982).
298 DeNoronha, F.; Baggs, R.; Schäfer, W.; Bolognesi, D. P.: Prevention of oncornavirus-induced sarcomas in cats by treatment with antiviral antibodies. Nature, Lond. *267:* 54 (1977).
299 Dent, P. B.; Fish, L. A.; White, L. G.; Good, R. A.: Chediak-Higashi syndrome. Observations on the nature of the associated malignancy. Lab. Invest. *15:* 1634 (1966).
300 DeVita, V. T.: Young, R. C.; Canellos, G. P.: Combination versus single agent

chemotherapy: a review of the basis for selection of drug treatment of cancer. Cancer 35: 98 (1975).

301 Dickneite, G.; Hofstaetter, T.; Schorlemmer, H. U.; Sedlacek, H. H.: Treatment of experimental chronic infection by an immunomodulating drug, bestatin. Int. J. Immunopharm. 4: 4 (1982).

302 Dickneite, G.; Kurrle, R.; Krajczewski, G.; Sedlacek, H. H.: Immunosuppressive action of aclacinomycin A and therapy of experimental graft versus host disease. Immunobiol. 162: 343 (1982).

303 Dickneite, G.; Johannsen, R.; Sedlacek, H. H.: The influence of bestatin, a small weight immunomodulator on chronic infection caused by salmonella typhimurium. Proc. 2nd Int. Symp. on Infections in the Immunocompromised Host, p. 155 London (1983).

304 Dickneite, G.; Kurrle, R.; Seiler, F. R.; Sedlacek, H. H.: Immunosuppression as a desired pharmacological effect. Behring Inst. Mitt. 74: 250 (1984).

305 Dickneite, G.; Schorlemmer, H.-U.; Sedlacek, H. H.: Chronic bacterial infection models for BRM screening. Behring Inst. Mitt. 74: 174 (1984).

306 Dickneite, G.; Kaspereit, F.; Sedlacek, H. H.: Stimulation of cell-mediated immunity by Bestatin leads to reduced bacterial persistance in experimental chronic Salmonella typhimurium infection. Infect. Immunity 44: 168 (1984).

307 Di Paola, M.; Angelini, L.; Lertolotti, A.; Colizza, S.: Host resistance in relation to survival in breast cancer. Br. med. J. iv: 268 (1974).

308 DiPaolo, J. A.: De Marinis, A. J.; Evans, E. H.; Doniger, J.: Expression of initiated and promoted stages of irradiation carcinogenesis in vitro. Cancer Lett. 14: 243 (1981).

309 DiPaolo, J. A.; Evans, C. H.; DeMarinis, A. J.; Doniger, J.: Phytohemagglutinin inhibits phorbol diester promotion of UV irradiation initiated transformation in Syrian hamster embryo cells. Int. J. Cancer 30: 781 (1982).

310 Dippold, W. G.; Lloyd, K. O.; Li, L. T. C.; Ikeda, H.; Oettgen, H. F.; Old, L. J.: Cell surface antigens of human malignant melanoma: Definition of six antigenic systems with mouse monoclonal antibodies. Proc. natn. Acad. Sci. USA 77: 6114 (1980).

311 Djeu, J. Y.; Heinbaugh, J. A.; Holden, H.; Herberman, R. B.: Augmentation of mouse natural killer activity by interferon and interferon inducers. J. Immun. 122: 175 (1979).

312 Djeu, J. Y.; Heinbaugh, J. A.; Holden, H. T.; Herberman, R. B.: Role of macrophages in the augmentation of mouse natural killer cell activity by poly I : C and interferon. J. Immun. 122: 182 (1979).

313 Djeu, J. Y.; Huang, K.-Y.; Herberman, R. B.: Augmentation of mouse natural killer activity and induction of interferon by tumor cells in vivo. J. exp. Med. 151: 781 (1980).

314 Doherty, P. C.; Bennink, J. R.: An examination of MHC restriction in the context of a minimal clonal abortion model for self tolerance. Scand. J. Immunol. 12: 271 (1980).

315 Doherty, P. C.; Schwartz, D. H.; Korngold, R.: B. Relationship between alloreactivity, self-MHC restriction, allo-MHC restriction of T cells from normal mice: frequency estimates: Self tolerance and MHC restriction. Behring Inst. Mitt. 70: 51 (1982).

316 Domzig, W.; Timonen, T. T.; Stadler, B. M.: Human natural killer (NK) cells produce interleukin-2 (IL-2). Proc. Am. Ass. Cancer Res. *22:* 309 (1981).
317 Donahoe, R. M.: Huang, K.: Neutralization of the phagocytosis enhancing activity of interferon preparations by anti-interferon serum. Infect. Immunity *7:* 501 (1973).
318 Donahoe, R. M.; Huang, K.: Interferon preparations enhance phagocytosis in vivo. Infect. Immunity *13:* 1250 (1976).
319 Doniach, D.; Hudwon, R. V.; Roitt, I. M.: Human auto-immune thyroiditis: clinical studies. Br. med. J. *i:* 365 (1960).
320 Doolittle, R. F.: Similar amino acid sequences: chance or common ancestry? Science *214:* 149 (1981).
321 Dorrington, K. J.: Properties of the Fc-receptor on macrophages and monocytes. Immunol. Commun. *5:* 263 (1976).
322 Douwes, F. R.; Spellmann, H. J.; Mross, K.; Wolfrum, D. J.: Immunodiagnostics of malignant disease. VI. Electrophoretic mobility test in malignant melanoma. Oncology *35:* 163 (1978).
323 Drake, W. P.; LeGendre, S. M.; Mardiney, M. R., Jr.: Depression of complement activity in three strains of mice after tumor transfer. Int. J. Cancer *11:* 719 (1973).
324 Droller, M. J.; Schneider, M. U.; Perlmann, P.: A possible role of prostaglandins in the inhibition of natural and antibody-dependent cell-mediated cytotoxicity against tumor cells. Cell. Imunol. *39:* 165 (1978).
325 Duff, R.; Rapp, F.: Reaction of serum from pregnant hamsters with surface of cells transformed by SV40. J. Immunol. *105:* 521 (1970).
326 Dunham, E. K.; Unanue, E. R.; Benacerraf, B.: Antigen binding and capping by lymphocytes of genetic nonresponder mice. J. exp. Med. *136:* 403 (1972).
327 Dupont, B. O.; Ballow, M.; Hansen, J. A.: Effect of transfer factor therapy on mixed lymphocyte culture reactivity. Proc. natn. Acad. Sci. USA *71:* 867 (1974).
328 Dvorak, A. M.; Connell, A. B.; Proppe, K.; Dvorak, H. J.: Immunologic rejection of mammary adenocarcinoma (TA3–ST) in C57Bl/6 mice: Participation of neutrophils and activated macrophages with fibrin formation. J. Immun. *120:* 1240 (1978).
329 Eaton, M. D.; Heller, J. A.; Scala, A. R.: Enhancement of lymphoma cell immunogenicity by infection with non-oncogenic virus. Cancer Res. *33:* 3293 (1973).
330 Eccles, S. A.: Studies on the effects of rat sarcomata on the migration of mononuclear phagocytes in vitro and in vivo; in James, The macrophage and cancer, p. 308 (James McBride and Stuart, Edinburgh 1977).
331 Eccles, S. A.; Alexander, P.: Macrophage content of tumors in relation to metastatic spread and host immune reaction. Nature *250:* 667 (1974).
332 Eccles, S. A.; Alexander, P.: Sequestration of macrophages in growing tumours and its effect on the immunological capacity of the host. Br. J. Cancer *30:* 42 (1974).
333 Eccles, S. A. et al.: Effect of cyclosporin A on the growth and spontaneous metastasis of syngeneic animal tumours. Br. J. Cancer *42:* 252 (1980).
334 Edelman, R.: Vaccine adjuvants. Rev. Infect. Dis. *2:* 370 (1980).
335 Edwards, P. A. W.; Foster, C. S.; McIlhinney, R. A. J.: Monoclonal antibodies to teratoma and breast. Transplant. Proc. *12:* 398 (1980).
336 Ehrlich, P.: On immunity with special reference to cell life. Proc. R. Soc. Lond. *66:* 424 (1900).

337 Ehrlich, P.: The relationship existing between chemical constitution, distribution and pharmacological action; in Ehrlich, Collected studies on immunity, ch. XXXIV, p. 441 (Wiley & Sons, New York 1906).
338 Ehrlich, P.: Experimentelle Carcinomstudien an Mäusen; in Himmelweit, Paul Ehrlich – Gesammelte Arbeiten, Band 2, Immunitätslehre und Krebsforschung, p. 493 (Springer, Berlin 1957).
339 Ehrlich, P.: Über den jetzigen Stand der Karzinomforschung; in Himmelweit, Paul Ehrlich – Gesammelte Arbeiten, Band 2, Immunitätslehre und Krebsforschung, p. 550 (Springer, Berlin 1957).
340 Eichelberg, D.; Schmutzler, W.: Pharmakologische Aspekte der Immunstimulanzien. Immun. Infekt. *11:* 109 (1983).
341 Eichmann, K.: Expression and function of idiotypes on lymphocytes. Adv. Immunol. *26:* 195 (1978).
342 Eichmann, K.; Rajewsky, K.: Induction of T and B cell immunity by anti-idiotypic antibody. Eur. J. Immunol. *5:* 661 (1975).
343 Eichmann, K.; Falk, I.; Rajewsky, K.: Recognition of idiotypes in lymphocyte interactions. II. Antigen-independent cooperation between T and B lymphocytes that possess similar and complementary idiotypes. Eur. J. Immunol. *8:* 853 (1978).
344 Eilber, F. R.; Morton, D. L.; Carmack Holmes, E.: Adjuvant immunotherapy with BCG in treatment of regional lymph node metastases from malignant melanoma. New Engl. J. Med. *294:* 237 (1976).
345 Eisenbach, L.; Segal, S.; Feldman, M.: MHC imbalance and metastatic spread in lewis lung carcinoma clones. Int. J. Cancer *32:* 113 (1983).
346 Eisenbarth, G. S.; Haynes, B. F.; Schroer, J. A.; Fauci, A. S.: Production of monoclonal antibodies reacting with peripheral blood mononuclear cell surface differentiation antigens. J. Immun. *124:* 1237 (1980).
347 Elgjo, K.; Degré, M.: Polyinosinic-polycytidylic acid in two-stage skin carcinogenesis. Effect on epidermal growth parameters and inferon induction in treated mice. J. natn. Cancer Inst. *51:* 171 (1973).
348 Ellner, J. J.; Rosenthal, A. S.: Quantitative and immunologic aspects of the handling of 2,4 dinitrophenyl guinea pig albumin by macrophages. J. Immunol. *114:* 1563 (1975).
349 Elston, C. W.; Bagshaw, K. D.: Cellular reaction in trophoblastic tumors. Br. J. Cancer *28:* 245 (1973).
350 Embleton, M. J.; Price, M. R.; Baldwin, R. W.: Demonstration and partial purification of common melanoma-associated antigen(s). Eur. J. Cancer *16:* 575 (1980).
351 Embleton, M. J.; Gunn, B.; Byers, V. S.: Antitumor reactions of monoclonal antibody against a human osteogenic-sarcoma cell line. Br. J. Cancer *43:* 582 (1981).
352 Engleman, E. G.; Charron, D. J.; Benike, C. J.; Stewart, G. H.: Ia antigen on peripheral blood mononuclear leukocytes in man. I. Expression, biosynthesis, and function of HLA-DR antigen non T-cells. J. exp. Med. *152:* 99 s (1980).
353 Enker, W. E.; Craft, K.; Wissler, R. W.: Active-specific immunotherapy with concanavalin-A-modified tumor cells. Transplant. Proc. *7:* suppl. 1, p. 489 (1975).
354 Epenetos, A. A.: Clinical results with regional antibody-guided irradiation. Advances in the applications of monoclonal antibodies in clinical oncology, University of London, Royal Postgraduate Medical School, 8th–10th May (1985).

355 Epenetos, A. A.; Courtenay-Luck, N.; Dhokia, B.; Snook, D.; Hooker, G.; Lavender, J. P.; Hemmingway, A.; Carr, D.; Papaharalambous, M.; Bosslet, K.; Buchegger, F.; Mach, J. P.: Antibody guided irradiation of hepatic metastases using intra-hepatically administered radiolabelled anti CEA antibodies with simultaneous and reversible hepatic blood stasis using biodegradable starch microspheres. Lancet (im Druck, 1986).
356 Eppinger-Helft, M.; Pavlovsky, S.; Hidalgo, G.; Sackmann, M. R.; Suarez, A.; Garay, G.; Russo, C.; Santos, M.; Macchi, A.; Lein, J.: Chemoimmunotherapy with cornyebacterium parvum in acute myelocytic leukemia. Cancer 45: 280 (1980).
357 Erb, P.; Feldmann, M.: The role of macrophages in the generation of T-helper cells. I. The requirement for macrophages in helper cell induction and characteristics of the macrophage T-cell interaction. Cell. Immunol. 19: 356 (1975).
358 Erb, P.; Feldmann, M.: The role of macrophages in the generation of T-helper cells. II. The genetic control of the macrophage-T-cell interaction for helper cell induction with soluble antigens. J. exp. Med. 142: 460 (1975).
359 Erb, P.; Feldmann, M.: Role of macrophages in in vitro induction of T helper cells. Nature, Lond. 254: 352 (1975).
360 Eremin, O.: NK cell activity in the blood, tumour-draining lymph nodes and primary tumours of women with mammary carcinoma; in Herberman, Natural cell-mediated immunity against tumors, p. 1011 (Academic Press, New York 1980).
361 Essex, M.; Grant, C. K.; Sliski, A. H.; Hardy, W. D., Jr.: Feline leukemia and immunological surveillance; in Chandra, NATO advances study institute series, series A: Life sciences, vol. 20, Antiviral mechanisms in the control neoplasia, p. 427 (Plenum Press, New York 1978).
362 Evans, R.: Macrophages in syngeneic animal tumors. Transplantation 14: 468 (1972).
363 Evans, C. H.: Lymphotoxin – An immunologic hormone with anticarcinogenic and antitumor activity. Cancer Immunol. Immunother. 12: 181 (1982).
364 Evans, C. H.: Lymphokines, homeostasis, and carcinogenesis. J. natn. Cancer Inst. 71: 253 (1983).
365 Evans, C. A.; Gorman, L. R.; Ito, Y.; Weiser, R. S.: Antitumor immunity in the shope papilloma-carcinoma complex of rabbits. I. Papilloma regression induced by homologous and autologous tissue vaccines. J. natn. Cancer Inst. 29: 277 (1962).
366 Evans, C. A.; Gorman, L. R.; Ito, Y.; Weiser, R. S.: Antitumor immunity in the shope papilloma-carcinoma complex of rabbits. II. Suppression of a transplanted carcinoma, V × 7, by homologous papilloma vaccine. J. natn. Cancer Inst. 29: 287 (1962).
367 Evans, R.; Booth, C. G.; Spencer, F.: Lack of correlation between in vivo rejection of syngeneic fibro-sarcomas and in vitro nonspecific macrophage cytotoxicity. Br. J. Cancer 38: 583 (1978).
368 Evans, C. H.; DiPaolo, J. A.: Lymphotoxin: An anticarcinogenic lymphokine as measured by inhibition of chemical carcinogen or ultraviolet-irradiation-induced transformation of Syrian hamster cells. Int. J. Cancer 27: 45 (1981).
369 Evans, C. H.; Heinbaugh, J. A.: Lymphotoxin cytotoxicity, a combination of cytolytic and cytostatic cellular responses. Immunopharmacol. 3: 347 (1981).
370 Evans, C. H., Di Paolo, J. A., Heinbaugh, J. A., DeMarinis, A. J.: Immunomodula-

tion of the lymphoresponsive phase of carcinogenesis: Mechanisms of natural immunity. J. natn. Cancer Inst. *69:* 737 (1982).

371 Evans, R. L.; Engleman, E. G.: Progress toward understanding self-tolerance. Semin. Hematol. *22:* 68 (1985).

372 Fagreus, A.; Espmark, A.: Detection of antigens in tissue culture with the aid of mixed hemadsorption. Acta path. microbiol. scand. suppl. 154, p. 258 (1962).

373 Fakhri, O.; Hobbs, J. R.: Studies of the rat immune response to plasmacytoma 5563 in C 3 H mice. Br. J. Cancer *24:* 853 (1970).

374 Fakhri, O.; Hobbs, J. R.: Overcoming some limiting factors in tumour immunotherapy. Br. J. Cancer *28:* 1 (1973).

375 Faraci, R. P.: In vitro demonstration of altered antigenicity of metastases from a primary methylcholanthrene-induced sarcoma. Surgery *76:* 469 (1974).

376 Farram, E.; Nelson, D. S.: Mechanism of action of mouse macrophages as antitumor effector cells: Role of arginase. Cell. Immunol. *55:* 283 (1980).

377 Fauve, R. M.; Hevin, B.; Jacob, H.: Anti-inflammatory effects of murine malignant cells. Proc. natn. Acad. Sci. USA *71:* 4052 (1974).

378 Fauve, R. M.; Hevin, M. B.: Toxic effects of tumour cells on macrophages; in James, The macrophage and cancer, p. 264 (James McBride and Stuart, Edinburgh 1977).

379 Fedyushin, M. P.: Antireticular cytotoxic serum in the treatment of cancer. Novyi khir. Arkh. *41:* 534 (1938).

380 Fefer, A.: Immunotherapy and chemotherapy of Moloney sarcoma virus-induced tumor in mice. Cancer Res. *29:* 2177 (1969).

381 Fefer, A.: Immunotherapy of primary Moloney sarcoma virus-induced tumors. Int. J. Cancer *5:* 327 (1970).

382 Feinstein, A.; Beale, D.: Models of immunoglobulins and antigen-antibody complexes; in Glynn, Steward, Immunochemistry: An advances textbook, p. 263 (John Wiley & Sons, Chichester 1977).

383 Feldman, J. D.: Immunological enhancement: A study of blocking antibodies. Adv. Immunol. *15:* 167 (1972).

384 Feldman, D. G.; Gross, L.: Electron microscopic study of spontaneous mammary carcinomas in cats and dogs: Virus-like particles in cat mammary carcinomas. Cancer Res. *31:* 1261 (1971).

385 Felgenhauer, K.: Immunological techniques following microelectrophoresis on polyacrylamide gel. Biochim. biophys. Acta *160:* 267 (1968).

386 Fenyö, E. M.; Klein, E.; Klein, G.: Selection of an immunoresistant Moloney lymphoma subline with decreased concentration of tumor-specific surface antigens. J. natn. Cancer Inst. *40:* 69 (1968).

387 Ferluga, J.; Schorlemmer, H. U.; Baptista, L. C.; Allison, A. C.: Cytolytic effects of the complement cleavage product, C3a. Br. J. Cancer *34:* 626 (1976).

388 Ferluga, J.; Schorlemmer, H. U.; Baptista, L. C.; Allison, A. C.: Production of the complement cleavage product, C3a, by activated macrophages and its tumorolytic effects. Clin. exp. Immunol. *31:* 512 (1978).

389 Festenstein, H.; Schmidt, W.: Variation in MHC antigenic profiles of tumor cells and its biological effects. Immunol. Rev. *60:* 85 (1981).

390 Fidler, I. J.: In vitro studies of cellular-mediated immunostimulation of tumor growth. J. natn. Cancer Inst. *50:* 1307 (1973).

391 Fidler, I. J.: Inhibition of pulmonary metastasis by intravenous injection of specifically activated macrophages. Cancer Res. *34:* 1074 (1974).
392 Fidler, I. J.: Biological behaviour of malignant melanoma cells correlated to their survival in vivo. Cancer Res. *35:* 218 (1975).
393 Fidler, I. J.: Recognition and destruction of target cells by tumoricidal macrophages. Israel J. med. Scis *14:* 177 (1978).
394 Fidler, I. J.: Tumor heterogeneity and the biology of cancer invasion and metastasis. Cancer Res. *38:* 2651 (1978).
395 Fidler, I. J.: The in situ induction of tumoricidal activity in alveolar macrophages by liposomes containing muramyl dipeptide is a thymus-independent process. J. Immunol. *127:* 1719 (1981).
396 Fidler, I. J.; Gersten, D. M.; Budmen, M. B.: Characterization in vivo and in vitro of tumor cells selected for resistance to syngeneic lymphocyte-mediated cytotoxicity. Cancer Res. *36:* 3160 (1976).
397 Fidler, I. J.; Roblin, R. O.; Poste, G.: In vitro tumoricidal activity of macrophages against virus-transformed lines with temperature-dependent transformed phenotypic characteristics. Cell. Immunol. *38:* 131 (1976).
398 Fidler, I. J.; Kripke, M. L.: Metastasis results from pre-existing variant cells within a malignant tumor. Science *197:* 893 (1977).
399 Fidler, I. J.; Barnes, Z.,; Fogler, W. E.; Kirsh, R.; Bugelski, P.; Poste, G.: Involvement of marcrophages in the eradication of established metastases following intravenous injection of liposomes containing macrophage activators. Cancer Res. *42:* 496 (1982).
400 Field, E. J.; Caspary, E. A.: Lymphocyte sensitization: An in vitro test for cancer? Lancet *ii:* 1337 (1970).
401 Finberg, R.; Burakoff, S.; Cantor, H.; Benacerraf, B.: Biological significance of alloreactivity: T cells stimulated by Sendai virus-coated syngeneic cells specifically lyse allogeneic target cells. Proc. natn. Acad. Sci. USA *75:* 5145 (1978).
402 Fink, M. A.; Smith, P.; Rothlauf, M. V.: Antibody production in Balb/c mice following injection of lyophilized tumor S621 in Freund's adjuvant. Proc. Soc. exp. Biol. Med. *90:* 590 (1955).
403 Fink, M. P.; Parker, C. W.; Shearer, W. T.: Antibody stimulation of tumour growth in T cell depleted mice. Nature, Lond. *255:* 404 (1975).
404 Finney, J. W.; Byers, E. H.; Wilson, R. H.: Studies in tumour auto-immunity. Cancer Res. *20:* 351 (1960).
405 Fisher, B.; Soliman, O.; Fisher, E. R.: Effect of ALS on parameters of tumor growth in a syngeneic tumor host system. Proc. Soc. exp. Biol. Med. *131:* 16 (1969).
406 Fisher, M. S.; Kripke, M. L.: Systemic alteration induced in mice by ultraviolet light irradiation and its relationship to ultraviolet carcinogenesis. Proc. natn. Acad. Sci. USA *74:* 1688 (1977).
407 Fisher, M. S.; Kripke, M. L.: Further studies on the tumor-specific suppressor cells induced by ultraviolet radiation. J. Immun. *121:* 1139 (1978).
408 Fisher, B.; Saffer, E. A.: Tumor cell cytotoxicity by granulocytes from peripheral blood of tumor-bearing mice. J. natn. Cancer Inst. *60:* 687 (1978).
409 Flax, M. H.: The action of anti-Ehrlich ascites tumor antibody. Cancer Res. *16:* 774 (1956).
410 Flechner, I.: The cure and concomitant immunization of mice bearing Ehrlich

ascites tumors by treatment with an antibody-alkylating agent complex. Eur. J. Cancer 9: 741 (1973).
411 Fogel, M.; Segal, S.; Gorelik, E.; Feldman, M.: Specific cytotoxic lymphocytes against syngeneic tumors are generated in culture in the presence of syngeneic, but not xenogeneic serum. Int. J. Cancer 22: 329 (1978).
412 Fogel, M.; Gorelik, E.; Segal, S.; Feldman, M.: Differences in cell surface antigens of tumor metastases and those of the local growth. J. natn. Cancer Inst. 62: 585 (1979).
413 Fogel, M. et al.: Metastatic potential severely altered by changes in tumor cell adhesiveness and cell-surface sialylation. J. exp. Med. 157: 371 (1983).
414 Foley, E. J.: Antigenic properties of methylcholanthrene-induced tumors in mice of the strain of origin. Cancer Res. 13: 835 (1953).
415 Foley, E. J.: Attempts to induce immunity against mammary adenocarcinoma in inbred mice. Cancer Res. 13: 578 (1953).
416 Fradelizi, D.; Gresser, I.: Interferon inhibits the generation of allospecific suppressor T lymphocytes. J. exp. Med. 155: 1610 (1982)
417 Franke, W. W.; Schiller, D. L.; Moll, R.; Winter, S.; Schmid, E.; Engelbrecht, I.; Denk, H.; Krepler, R.; Platzer, B.: Diversity of cytokeratins. Differentiation specific expression of cytokeratin polypeptides in epithelial cells and tissues. J. molec. Biol. 153: 933 (1981).
418 Frankel, A. E.; Rouse, R. V.; Herzenberg, L. A.: Human prostate specific and shared differentiation antigens defined by monoclonal antibodies. Proc. natn. Acad. Sci. USA 79: 903 (1982).
419 Fraser, K. B.: The formation of antibody. A study of the relationship between a normal and an immune haemagglutinin. J. Path. Bact. 70: 13 (1955).
420 Freedman, R. S.; Bowen, J. M.; Herson, J.; Wharton, J. T.; Rutledge, F. N.; Hamberger, A. D.: Virus-modified homologous tumor-cell extract in the treatment of vulvar carcinoma. Cancer Immunol. Immunother. 8: 33 (1980).
421 Freedman, V. H.; Calvelli, T. A.; Silagi, S.; Silverstein, S. C.: Macrophages elicited with heat-killed Bacillus Calmette-Guerin protect C57Bl/6J mice against a syngeneic melanoma. J. exp. Med. 15: 657 (1980).
422 Friedenreich, V.: Untersuchungen über das von O. Thomsen beschriebene vermehrungsfähige Agens als Veränderer des isoagglutinatorischen Verhaltens der roten Blutkörperchen. Z. ImmunForsch. 55: 84 (1928).
423 Friedenreich, V.: Die serologische Auffassung des Thomschen Blutkörperchenrezeptors. Acta path. microbiol. scand., suppl. V, p. 68 (1930).
424 Fritze, D.; Kern, D. H.; Dorgemuller, C. R.; Pilch, Y. H.: Production of antisera with specificity for malignant melanoma and human fetal skin. Cancer Res. 36: 458 (1976).
425 Fritze, D.; Pilch, Y. H.; Kern, D. H.: Induktion spezifischer Tumor-Immunität mit anti-tumor Immun RNA. Teil II: Therapeutische Konsequenzen. Lab. med. 4: 230 (1980).
426 Fritze, D.; Becher, R.; Massner, B.; Kaufmann, M.; Bruntsch, U.; Gallmeier, W. M.; Mayr, A. C.; Drings, P.; Abel, U.; Edler, L.; Jungi, W. F.; Queißer, W.; Senn, H. J.: A randomized study of combination chemotherapy (VAC-FMC) with or without immunostimulation by Corynebacterium parvum in metastatic breast cancer. Klin. Wschr. 60: 593 (1982).

427 Fröland, S. S.: Binding of sheep erythrocytes to human lymphocytes. A probable marker of T lymphocytes. Scand. J. Immunol. *1:* 269 (1972).
428 Fröland, S. S.; Natvig, J. B.: Class, subclass and allelic exclusion of membrane-bound Ig of human B-lymphocytes. J. exp. Med. *136:* 409 (1972).
429 Fröland, S. S.; Wislöff, F.; Michaelson, T. E.: Human lymphocytes with receptors for IgG: a population of cells distinct from T and B lymphocytes. Eur. J. Immunol. *4:* 302 (1974).
430 Frost, P.; Sanderson, C. J.: Tumor immunoprophylaxis in mice using glutaraldehyde-treated syngeneic tumor cells. Cancer Res. *35:* 2646 (1975).
431 Frost, P.; Edwards, A.; Sanderson, C.: The use of glutaraldehyde fixation for the study of the immune response to syngeneic tumor antigen. Ann. N. Y. Acad. Sci. *276:* 91 (1976).
432 Fry, W. A. et al.: Lung cancer patients autoimmune responses to Thomsen-Friedenreich (T)-antigen: Diagnostic utility. Klin. Wschr. *61:* 817 (1983).
433 Frye, L. D.; Friou, G. H.: Inhibition of mammalian cytotoxic cells by phosphatidylcholine and its analogue. Nature, Lond. *258:* 333 (1975).
434 Fuhrer, J. P.; Evans, C. H.: The anticarcinogenic and tumor growth inhibitory activities of lymphotoxin are associated with altered membrane glycoprotein synthesis. Cancer Lett. *19:* 283 (1983).
435 Fuhrer, J. P.; Evans, C. H.: Human lymphotoxin but not interferon differentially modulates glucosaminyl fucosyl glycoprotein synthesis in human cells. Fed. Proc. *42:* 681 (1983).
436 Fujimoto, S.; Greene, M.; Sehon, A. H.: Regulation of the immune response to tumor antigens. I. Immunosuppressor cells in tumor-bearing hosts. J. Immunol. *116:* 791 (1976).
437 Fujimoto, S.; Greene, M.; Sehon, A. H.: Regulation of the immune response to tumor antigens. II. The nature of immunosuppressor cells in tumor-bearing hosts. J. Immunol. *116:* 800 (1976).
438 Fujiwara, M.; Natata, T.: Induction of tumour immunity with tumour cells treated extract of garlic (allium sativum). Nature, Lond. *216:* 83 (1967).
439 Fukushima, M.; Machida, S.; Hokama, A.; Kojika, M.; Nishikawa, T.; Kikuchi, A.; Ishikawa, Y.: Passive transfer of the resistance to tumor with RNA. Tohoku J. exp. Med. *112:* 115 (1974).
440 Fukushima, M.; Machida, S.; Nishikama, T.; Ishikawa, Y.: Antitumor effect of allogeneic lymphocytes sensitized with tumor-specific immune RNA on human cancer cells. Cell. mol. Biol. *25:* 39 (1979).
441 Fukushima, M.; Colmerauer, M. E. M.; Nayak, S. K.; Koziol, J. A.; Pilch, Y. H.: Immunotherapy of a murine colon cancer with syngeneic spleen cells, immune RNA and tumor antigen. Int. J. Cancer *29:* 107 (1982).
442 Fukushima, K.; Hirota, M.; Terasaki, P. I.; Wakisaka, A.; Togashi, H.; Chia, D.; Suyama, N.; Fukushi, Y.; Nudelman, E.; Hakomori, S.-I.: Characterization of sialosylated Lewis as a new tumor-associated antigen. Cancer Res. *44:* 5279 (1984).
443 Gale, R. P.: Advances in the treatment of acute myelogenous leukemia. New Engl. J. Med. *300:* 1189 (1979).
444 Gale, R. P.; Zighelboim, T.; Ossorio, C.; Fahey, T. L.: Immunologically specific tumor cell destruction by human polymorphonuclear leukocytes. Clin. Res. *180:* 180 A (1974).

445 Galili, U.; Vanky, F.; Rodriguez, L.; Klein, E.: Activated T lymphocytes within human solid tumors. Cancer Immunol. Immunother. *6:* 129 (1979).
446 Gallego, J.; Price, M. R.: Monoclonal antibody-drug conjugates: A new approach for cancer therapy. Drugs of today *21:* 511 (1985).
447 Gallin, E. K.; Seligmann, B.; Gallin, J. I.: Alteration of macrophage and monocyte membrane potential by chemotactic factors; in Van Furth, Mononuclear phagocytes: Functional aspects, part II, p. 19 (Martinus Nijhoff Publishers, The Hague, Boston, London 1980).
448 Galton, D. A. G.; Kay, H. E. M.; Reizenstein, P.; Penchansky, M.; Vogler, W. R.; Whittacker, J. A.: Infection and second-remission rates in patients having immunotherapy for acute myeloid leukemia. Lancet *i:* 973 (1977).
449 Garnis, S.: Surface properties of small lymphocytes appearing in the Ehrlich ascites tumor and host spleens. Anat. Rec. *184:* 408 (1976).
450 Garrido, F.; Festenstein, H.; Schirrmacher, V.: Further evidence for derepression of H-2 and Ia-like specificities of foreign haplotypes in mouse tumour cell lines. Nature *261:* 705 (1976).
451 Garrido, F.; Schirrmacher, V.; Festenstein, N. H.: H-2 like specificities of foreign haplotypes appearing in a mouse sarcoma after vaccine virus infection. Nature, Lond. *259:* 228 (1976).
452 Gateley, M. K.; Mayer, M. M.; Henney, C. S.: Effect of anti-lymphotoxin on cell-mediated cytotoxicity. Evidence for two pathways, one involving lymphotoxin and the other requiring intimate contact between the plasma membranes of killer and target cells. Cell. Immunol. *27:* 82 (1976).
453 Gautam, S.; Aikat, B. K.: Role of fetal antigens in tumor immunity. Eur. J. Cancer *13:* 71 (1977).
454 Geczy, A.; de Weck, A. L.: Molecular basis of T cell dependent genetic control of the immune response in the guinea pig; in Kallós, Progress in allergy, vol. 22, p. 147 (Karger, Basel 1977).
455 Gee, T. S.; Dowling, M. D.; Cunningham, I.: Evaluation of Pseudomonas aeruginosa vaccine for prolongation of remissions in adults with acute nonlymphoblastic leukemia trated with the L-12 protocol: A preliminary report; in Terry, Windorst, Immunotherapy of cancer: Present status of trials in man, p. 415 (Raven Press, New York 1978).
456 Gelboin, H. V.; Levy, H. B.: Polyinosinic-polycytidylic acid inhibits chemically induced tumorigenesis in mouse skin. Science *167:* 205 (1970).
457 Gemsa, D.; Seitz, M.; Kramer, W.; Till, G.; Resch, K.: The effects of phagocytosis, dextran sulfate, and cell damage on PGE1 sensitivity and PGE1 production of macrophages. J. Immun. *120:* 1187 (1978).
458 Gerlier, D.; Price, M. R.; Bisby, R. H.; Baldwin, R. W.: Modification of cell cholesterol content of rat tumour cells: Effect upon their tumorigenicity and immunogenicity. Cancer Immunol. Immunother. *14:* 27 (1982).
459 Gershon, R. K.: T cell control of antibody production. Contemp. Top. Immunobiol. *3:* 1 (1974).
460 Gershon, R. K.; Mokyr, M. B.; Mitchell, M. S.: Activation of suppressor T cells by tumour cells and specific antibody. Nature *250:* 594 (1974).
461 Geshwin, M. E.; Steinberg, A. D.: Loss of suppressor function as a cause of lymphoid malignancy. Lancet *ii:* 1174 (1973).

462 Gerson, J. M.: Systemic and in situ natural killer activity in tumor-bearing mice and patients with cancer; in Herberman, Natural cell-mediated immunity against tumors, p. 1047 (Academic Press, New York 1980).

463 Gerson, J. M., Varesio, L., Herberman, R. B.: Systemic and in situ natural killer and suppressor cell activities in mice bearing progressively growing murine sarcoma virus-induced tumors. Int. J. Cancer 27: 243 (1981).

464 Gery, I.; Gershon, R. K.; Waksman, B. H.: Potentiation of the thymus derived lymphocyte response to mitogens. I. The responding cell. J. exp. Med. 136: 128 (1972).

465 Ghose, T.; Norvell, S. T.; Guclu, A.; Cameron, D.; Bodurtha, A.; MacDonald, A. S.: Immunochemotherapy of cancer with chlorambucil-carrying antibody. Br. med. J. iii: 495 (1972).

466 Ghose, T.; Guclu, A.; Tsai, J.; Mammen, M.; Norvell, S. T.: Immunoprophylaxis and imunotherapy of EL4 lymphoma. Eur. J. Cancer 13: 925 (1977).

467 Ghose, T.; Blair, A. H.: Antibody-linked cytotoxic agents in the treatment of cancer: Current status and future prospects. J. natn. Cancer Inst. 61: 657 (1978).

468 Gidlund, M.; Örn, A.; Wigzell, H.; Senik, A.; Gresser, I.: Enhanced NK cell activity in mice injected with interferon and interferon inducers. Nature, Lond. 223: 259 (1978).

469 Gillespie, G. Y.; Hansern, C. B.; Haskins, R. G.; Russell, S. W.: Inflammatory cells in solid murine neoplasms. IV. Cytolytic T lymphocytes isolated from regressing or progressing Moloney sarcomas. J. Immun. 119: 564 (1977).

470 Gilette, R. W.; Fox, A.: The effect of T lymphocyte deficiency on tumor induction and growth. Cell. Immunol. 19: 328 (1975).

471 Gilliland, D. G.; Steplewski, Z.; Callier, R. J.: Antibody-directed cytotoxic agents: Use of monoclonal antibody to direct the action of toxin A chains to colorectal carcinoma cells. Proc. natn. Acad. Sci. USA 77: 4539 (1980).

472 Girard, J. P.; Fernandes, B.: Studies on the mitogenic activity of trypsin, pronase and neuraminidase on human peripheral blood lymphocytes. Eur. J. clin. Invest. 6: 347 (1976).

473 Girardi, A. J.: Prevention of SV40 virus oncogenesis in hamsters. I. Tumor resistance induced by human cells transformed by SV40. Proc. natn. Acad. Sci. USA 54: 445 (1965).

474 Glaser, M.: Regulation of specific cell-mediated cytotoxic response against SV40-induced tumor associated antigens by depletion of suppressor T cells with cyclophosphamide in mice. J. exp. Med. 149: 774 (1979).

475 Glaser, M.; Djeu, J. Y.; Kirchner, H.; Herberman, R. B.: Augmentation of cell-mediated cytotoxicity against syngeneic gross virus-induced lymphoma in rats by phytohemagglutinin and endotoxin. J. Immun. 116: 1512 (1976).

476 Glennie, M. J.; Stevenson, G. T.: Univalent antibodies kill tumour cells in vitro and in vivo. Nature 295: 712 (1982).

477 Godleski, J. J.; Lee, R. E.; Leighton, J.: Studies on the role of polymorphonuclear leukocytes in neoplastic disease with the chick embryo and walker carcinosarcoma 256 in vivo and in vitro. Cancer Res. 30: 1986 (1970).

478 Götze, O.; Bianco, C.; Sundsmo, J. S.; Cohn, Z. A.: The stimulation of mononuclear phagocytes by components of the classical and the alternative pathways of complement activation; in van Furth, Mononuclear phagocytes: Functional aspects, part II, p. 55 (Martinus Nijhoff Publishers, The Hague, Boston, London 1980).

479 Gold, P.: Embryonic origin of human tumor-specific antigens; in van Duuren, Progress in experimental tumor research, vol. 14, p. 43 (Karger, Basel 1971).
480 Gold, P.; Freedman, S. O.: Specific carcinoembryonic antigens of the human digestive system. J. exp. Med. *102:* 467 (1965).
481 Goldberg, B.; Green, H.: The cytotoxicity of immune gamma globulin and complement in Krebs ascites tumor cells. I. Ultrastructural studies. J. exp. Med. *109:* 505 (1959).
482 Goldenberg, D. M.: An introduction to the radioimmunodetection of cancer. Cancer Res. *40:* 2957 (1980).
483 Goldenberg, D. M.; Pavia, R. A.; Tsao, M. C.: In vivo hybridization of human tumor and normal hamster cells. Nature *250:* 649 (1974).
484 Goldfarb, R. H.; Herberman, R. B.: Natural killer cell reactivity: Regulatory interactions among phorbol ester, interferon, cholera toxin, and retinoic acid. J. Immun. *126:* 2129 (1981).
485 Goldfeder, A.: Relation between radiation effects and cell viability as indicated by induced resistance to transplanted tumors. Radiology *39:* 426 (1942).
486 Golstein, P.; Denizot, F.; Samuel, D.; Tayler, R.; Rubin, B.: Xenogeneic serum-induced interleukin(s) that trigger the differentiation of precursor into cytolytic T cells. Behring Inst. Mitt. *67:* 80 (1980).
487 Gomard, E.; Duprez, V.; Henin, Y.; Levy, J. P.: H-2 region product determinant in immune cytolysis of syngeneic tumour cells by anti-MSV T lymphocytes. Nature *250:* 707 (1976).
488 Gonwa, T. A.; Peterlin, B. M.; Stobo, J. D.: Human Ir genes: Structure and function. Adv. Immunol. *34:* 71 (1983).
489 Gonwa, T. A.; Picker, L. J.; Raff, H. V.; Goyert, S. M.; Silver, J.; Stabo, J. D.: Antigen-presenting capabilities of human monocytes correlate with their expression of HLA-DS, an Ia determinant distinct from HLA-DR. J. Immun. *130:* 706 (1983).
490 Good, R. A.; Dalmasso, A. P.; Martinez, C.; Archer, O. K.; Pierce, J. C.; Papermaster, B. W.: The role of the thymus in development of immunologic capacity in rabbits and mice. J. exp. Med. *116:* 773 (1962).
491 Goodwin, J. S.; Messner, R. P.; Bankhurst, A. D.: Prostaglandin-producing suppressor cells in Hodgkin's disease. New Engl. J. Med. *297:* 963 (1977).
492 Gorczynski, R. M.: Immunity to murine sarcoma virus induced tumors. I. Specific T lymphocytes active in macrophage migration inhibition and lymphocyte transformation. J. Immun. *112:* 1815 (1974).
493 Gorczynski, R. M.: Immunity to murine sarcoma virus-induced tumors. II. Suppression of T cell-mediated immunity by cells from progressor animals. J. Immun. *112:* 1826 (1974).
494 Gorczynski, R. M.; Kilburn, D. G.; Knight, R. A.; Norbury, C.; Parker, D. C.; Smith, J. B.: Nonspecific and specific immunosuppression in tumour-bearing mice by soluble immune complexes. Nature *254:* 141 (1975).
495 Gordon, S.: Macrophage neutral proteinases and chronic inflammation. Ann. N. Y. Acad. Sci. *278:* 176 (1976).
496 Gordon, S.: Lysozyme and plasminogen activator – constitutive and induced secretory products of mononuclear phagocytes; in van Furth, Mononuclear phagocytes, p. 1273 (Nijhoff, The Hague 1980).

497 Gordon, S.; Todd, J.; Cohn, Z. A.: In vitro synthesis and secretion of lysozyme by mononuclear phagocytes. J. exp. Med. *139:* 1228 (1974).
498 Gordon, D.; Bray, M. A.; Morley, J.: Control of lymphokine secretion by prostaglandins. Nature, Lond. *262:* 401 (1976).
499 Gorer, P. A.: The genetic and antigenetic basis of tumour transplantation. J. Path. Bact. *44:* 691 (1937).
500 Gorer, P. A.: The antigenetic basis of tumour transplantation. J. Path. Bact. *47:* 231 (1938).
501 Gorer, P. A.; O'Gorman, P.: The cytotoxic activity of isoantibodies in mice. Transplant. Bull. *3:* 142 (1956).
502 Goutner, A.; Schwarzenberg, L.; Mathe, G.: Phase I study of azimexon in immunodepressed cancer patients; in Terry, Rosenberg, Immunotherapy of human cancer, p. 471 (Elsevier, New York 1981).
503 Gown, A. M.; Vogel, A. M.: Monoclonal antibodies to intermediate filament proteins of human cells: Unique and cross-reacting antibodies. J. cell. Biol. *95:* 414 (1982).
504 Grabar, P.; Williams, C. A.: Méthode permettant l'étude conjugée des propriétés électrophorétiques et immunologiques d'un mélange de protéines. Application au sérum sanguin. Biochim. biophys. Acta *10:* 193 (1953).
505 Graham, J. B.; Graham, R. M.: Autogenous vaccine in cancer patients. Surgery Gynec. Obstet. *114:* 1 (1962).
506 Granger, G. A.; Shacks, S. J.; Williams, T. W.; Kolb, W. P.: Lymphocyte in vitro cytotoxicity, specific release of lymphotoxin-like materials from tuberculin-sensitive lymphoid cells. Nature *221:* 1155 (1969).
507 Grant, C. A.; Miller, J. F. A. P.: Effect of neonatal thymectomy on the induction of sarcomata in C57Bl mice. Nature *205:* 1124 (1965).
508 Grant, C. K.; Harris, D.; Essex, M. E.; Pickard, D. K.; Hardy, W. D., Jr.; de Noronha, F.: Protection of cats against feline leukemia virus-positive and virus-negative tumors by complement-dependent antibody. J. natn. Cancer Inst. *64:* 1527 (1980).
509 Greaves, M. F.: Analysis of the clinical and biological significance of lymphoid phenotypes in acute leukemia. Cancer Res. *41:* 4752 (1981).
510 Green, I.; Paul, W. E.; Benacerraf, B.: A study of the passive transfer of delayed hypersensitivity to DNP-poly-L-lysine and DNP-GL in responder and nonresponder guinea pigs. J. exp. Med. *126:* 959 (1967).
511 Green, A. A.; Pratt, C.; Webster, R. G.; Smith, K.: Immunotherapy of osteosarcoma patients with virus-modified tumor cells. Ann. N. Y. Acad. Sci. *277:* 396 (1976).
512 Green, S.; Dobrjansky, A.; Carswell, E. A.; Kassel, R. L.; Old, L. J.; Fiore, N.; Schwartz, M. K.: Partial purification of a serum factor that causes necrosis of tumors. Proc. natn. Acad. Sci. USA *73:* 381 (1976).
513 Green, S.; Dobrjanski, A.; Chiasson, M. A.; Carswell, E.; Schwartz, M. K.; Old, L. J.: Corynebacterium parvum as the priming agent in the production of tumor necrosis factor in the mouse. J. natn. Cancer Inst. *59:* 1519 (1977).
514 Greene, M. I.; Dorf, M. E.; Pierres, M.: Reduction of syngeneic tumour growth by an anti-I-J alloantiserum. Proc. natn. Acad. Sci. USA *74:* 5118 (1977).
515 Greene, M. I.; Fujimoto, S.; Sehon, A. H.: Regulation of the immune response to tumor antigens. III. Characterization of thymic suppressor factor(s) produced by tumor-bearing hosts. J. Immun. *119:* 757 (1977).

516 Greene, M. I.; Perry, L. L.; Benacerraf, B.: Regulation of the immune response to tumor antigen. Am. J. Path. *95:* 159 (1979).
517 Greiner, J. W.; Hand, P. H.; Noguchi, P.; Fisher, P. B.; Pestka, S.; Schlom, J.: Enhanced expression of surface tumor-associated antigens on human breast and colon tumor cells after recombinant human leukocytes α-Interferon treatment. Cancer Res. *44:* 3208 (1984).
518 Gresser, I.; Brouty-Boyé, D.; Thomas, M. G.; Mazieira-Coelho, A.: Interferon and cell division. I. Inhibition of the multiplication of mouse leukemia L1210 cells in vitro by interferon preparations. Proc. natn. Acad. Sci. USA *66:* 1052 (1970).
519 Gresser, I.; Vignaux, F.; Maury, C.; Lindahl, P.: Factor(s) from Ehrlich ascites cells responsible for delayed rejection of skin allografts in mice and its assay on lymphocytes in vitro. Proc. Soc. exp. Biol. Med. *149:* 83 (1975).
520 Griffith, I. P.; Crook, N. E.; White, D. O.: Protection of mice against cancer by immunization with membranes but not purified virions from virus infected cancer cells. Br. J. Cancer *31:* 603 (1975).
521 Griscelli, C.; Durandy, A.; Guy-Grand, D.; Daguillard, F.; Herzog, C.; Prunieras, M.: A syndrome associating partial albinism and immunodeficiency. Am. J. Med. *65:* 691 (1978).
522 Gross, L.: Intradermal immunization of C3H mice against a sarcoma that originated in an animal of the same line. Cancer Res. *3:* 326 (1943).
523 Gross, L.: The specificity of acquired tumor immunity. J. Immun. *50:* 91 (1945).
524 Grosser, N.; Mari, J. H.; Proctor, J. W.; Thomsen, D. M. P.: Tube leukocyte adherence inhibition assay for the detection of anti-tumor immunity. I. Monocyte is the reactive cell. Int. J. Cancer *18:* 39 (1976).
525 Grosser, N.; Thomson, D. M. P.: Tube leukocyte (monocyte) adherence, inhibition assay for the detection of antitumor immunity. III. «Blockade» of monocyte reactivity by excess free antigen and immune complexes in advanced cancer patients. Int. J. Cancer *18:* 58 (1976).
526 Gupta, R. K.; Silver, H. K. B.; Reisfeld, R. A.; Morton, D. L.: Isolation and immunochemical characterization of antibodies from the sera of cancer patients which are reactive against human melanoma cell membranes by affinity chromatography. Cancer Res. *39:* 1683 (1979).
527 Gutschank, S.; Rothauge, C. F.; Kraushaar, J.; Gutschank, W.; Sedlacek, H. H.: Das entgleiste metastasierende Prostatakarzinom. Münch. med. Wschr. *4:* 133 (1981).
528 Habel, K.: Resistance of polyoma virus immune animals to transplanted polyoma tumors. Proc. Soc. exp. Biol. Med. *106:* 722 (1961).
529 Habel, K.: Virus tumor antigens: Specific fingerprints? Cancer Res. *26:* 2018 (1966).
530 Habel, K.; Jensen, F.; Pagano, J. S.; Koprowski, H.: Specific complement-fixing tumor antigen in SV40-transformed human cells. Proc. Soc. exp. Biol. Med. *118:* 4 (1965).
531 Häkkinen, I.; Halonen, P.: Induction of tumor immunity in mice with antigens prepared form influenza and vesicular stomatitis virus grown in suspension culture of Ehrlich ascites cells. J. natn. Cancer Inst. *46:* 1161 (1971).
532 Hakamori, S. I.: Glycolipid changes associated with malignant transformation. In: 22nd Colloquium of the Gesellschaft für Biologische Chemie, 15.– 7. April 1971,

Mosbach/Baden; in Wallach, Fischer, The dynamic structure of cell membranes, p. 65 (Springer, Heidelberg, New York 1971).
533 Hakomori, S.; Murakami, W. T.: Glycolipids of hamster fibroblasts and derived malignant-transformed cell lines. Proc. natn. Acad. Sci. USA 59: 254 (1968).
534 Hakamori, S.; Kannagi, R.: Glycosphingolipids as tumor-associated and differentiation markers. J. natn. Cancer Inst. 71: 231 (1983).
535 Hakomori, S.-I.: Philipp Levine Award Lecture: Blood group glycolipid antigens and their modification as human cancer antigens. Am. J. clin. Path. 82: 635 (1984).
536 Hakomori, S.-I.: Tumor-associated carbohydrate antigens. Annu. Rev. Immunol. 2: 103 (1984).
537 Haller, O. et al.: Role of non-conventional natural killer cells in resistance against syngeneic tumour cells in vivo. Nature, Lond. 270: 609 (1977).
538 Halliday, W. J.: Macrophage migration inhibition with mouse tumor antigens: Properties of serum and peritoneal cells during tumor growth and after tumor loss. Cell. Immunol. 3: 113 (1972).
539 Halliday, W. J.; Miller, S.: Leukocyte-adherence inhibition: A simple test for cell-mediating tumour immunity and serum blocking factors. Int. J. Cancer 9: 477 (1972).
540 Halliday, W. J.; Maluish, A. E.; Isbister, W. H.: Detection of anti-tumour cell-mediated immunity and serum blocking factors in cancer patients by the leukocyte adherence inhibition test. Br. J. Cancer 29: 31 (1974).
541 Halliday, W. J.; Maluish, A. E.; Miller, S.: Blocking and unblocking of cell-mediated anti-tumor immunity in mice, as detected by the leukocyte adherence inhibition test. Cell. Immunol. 10: 467 (1974).
542 Halliday, W. J.; Maluish, A. E.; Little, J. H.; Davis, N. C.: Leukocyte adherence inhibition and specific immunoreactivity in malignant melanoma. Int. J. Cancer 16: 645 (1975).
543 Halpern, B.: Corynebacterium parvum: Applications in experimental and clinical oncology (Plenum Press, New York 1975).
544 Halpern, B.; Prevot, A. R.; Biozzi, G.; Stiffel, C.; Mouton, D.; Morard, J. C.; Bouthilier, Y.; Decreusefond, C.: Stimulation de l'activité phagocytaire du système reticuloendothelial provoqué par Corynebact. parvum. J. reticuloendoth. Soc. 1: 77 (1964).
545 Hamburg, S. I.; Cassell, G. H.; Rabinovitch, M.: Relationship between enhanced macrophage phagocytic activity and the induction by Newcastle disease virus in mice. J. Immun. 124: 1360 (1980).
546 Hammond, W. G.; Fisher, J. C. Rolley, R. T.: Tumor-specific transplantation immunity to spontaneous mouse tumors. Surgery 62: 124 (1967).
547 Hanna, M. G., Jr.: Active specific immunotherapy of residual micrometastasis: A comparison of postoperative treatment with BCG tumor cell vaccine to preoperative intratumoral BCG injection; in Terry, Yamamura, Immunobiology and immunotherapy, p.415 (Elsevier/North-Holland, New York 1979).
548 Hanna, M. G., Jr.; Peters, L. C.: Immunotherapy of established micrometastases with Bacillus Calmette-Guérin tumor cell vaccine. Cancer Res. 38: 204 (1978).
549 Hanna, M. G., Jr.; Brandhorst, J. S.; Peters, L. C.: Active specific immunotherapy of residual micrometastasis. An evaluation of sources, doses and ratios of BCG with tumor cells. Cancer Immunol. Immunother. 7: 165 (1979).

550 Hanna, M. G., Jr.; Peters, L. C.: Specific immunotherapy of established visceral micrometastases by BCG-tumor cell vaccine alone or as an adjunct to surgery. Cancer 42: 2613 (1979).
551 Hansen, T. H.; Ozato, K.; Sachs, D. H.: Heterogeneity of H-2D region associated genes and gene products. Adv. Immunol. 34: 39 (1983).
552 Hansson, M.; Kärre, K.; Kiessling, R.; Roder, J.; Andersson, B.; Häyry, P.: Natural NK-cell targets in the mouse thymus: Characteristics of the sensitive cell population. J. Immunol. 123: 765 (1979).
553 Haranka, K.; Satomi, N.; Sakurai, A.: Antitumor activity of murine tumor necrosis factor (TNF) against transplanted murine tumors and heterotransplanted human tumors in nude mice. Int. J. Cancer 34: 263 (1984).
554 Harder, F.-H.; McKhann, C. F.: Demonstration of cellular antigens on sarcoma cells by an indirect ^{125}J-labeled antibody technique. J. natn. Cancer. Inst. 40: 231 (1968).
555 Hardy, W. D., Jr.; Hess, P. W.; MacEwen, E. G.; Hayes, A. A.; Kassel, R. L.; Day, N. K.; Old, L. J.: Treatment of feline lymphosarcoma with feline blood constituents; in Clemmesen, Yohn, Comparative leukemia research. Bibl. Haemat., vol. 43, p. 518 (Karger, Basel 1976).
556 Harmon, R. C.; Clark, E.; O'Toole, C.; Wicker, L.: Resistance of H-2 heterozygous mice to parental tumors. I. Hybrid resistance and natural cytotoxicity to EL-4 are controlled by the H-2D-Hh-1 region. Immunogenetics 4: 601 (1977).
557 Harris, J.: Immune deficiency states associated with malignant disease in man. Med. Clins N. Am. 56: 501 (1972).
558 Harris, R.; Zuhrie, S. R.; Freeman, C. B.; Tayler, G. M.; MacIvery, J. E.; Geary, C. G.; Delamore, I. W.; Hull, P. J.; Tooth, J. A.: Active immunotherapy in acute myelogenous leukaemia and the induction of second and subsequent remissions. Br. J. Cancer 37: 282 (1978).
559 Harthus, H. P.; Johannscon, R.; Ax, W.: Lymphocyte sensitization in tumor-bearing rats: EMT versus MLTC. Immunobiology 156: 268 (1979).
560 Harthus, H.-P.; Ax, W.: Electrophoretic mobility test (EMT): Studies on lymphocyte response and mechanism of the test using a rat tumor model. Immunobiology 158: 151 (1981).
561 Harvath, L.; Amirault, H. J.; Anderson, B. R.: Chemilluminescence of human and canine polymorphonuclear leukocytes in the absence of phagocytosis. J. clin. Invest. 59: 1145 (1978).
562 Haskill, R. S.; Yamamura, Y.; Radov, L.: Host response within solid tumors: Non-thymus-derived specific cytotoxic cells within a murine mammary adenocarcinoma. Int. J. Cancer 16: 798 (1975).
563 Hauschka, T. S.; Weiss, L.; Holdrigde, H.: Karyotypic and surface features of murine TA3 carcinoma cells during immunoselection in mice and rats. J. natn. Cancer Inst. 47: 343 (1971).
564 Hauser, P.; Vaes, G.: Degradation of cartilage proteoglycans by a neutral proteinase secreted by rabbit bone-marrow macrophages in culture. Biochem. J. 172: 275 (1978).
565 Hausman, M. S.; Brosman, S.; Synderman, R.; Mickey, M. R.; Fahey, J.: Defective monocyte function in patients with genitourinary tract carcinoma. J. natn. Cancer Inst. 55: 1047 (1975).

566 Hayes, H. M., Jr.: The comparative epidemiology of selected neoplasms between dogs, cats and humans. A review. Eur. J. Cancer *14:* 1299 (1978).
567 Haynes, B. F.: Human T lymphocyte antigens as defined by monoclonal antibodies. Immunol. Rev. *57:* 127 (1981).
568 Haynes, B. F.; Miller, S. E.; Palker, T. J.; Moore, J. O.; Dunn, P. H.; Bolognesi, D. P.; Metzgar, R. S.: Identification of human T cell leukemia virus in a Japanese patient with adult T cell leukemia and cutaneous lymphomatous vasculitis. Proc. natn. Acad. Sci. USA *80:* 2054 (1983).
569 Haywood, G. R.; McKhann, C. F.: Antigenic specificities on murine sarcoma cells. Reciprocal relationship between normal transplantation antigens (H-2) and tumor-specific immunogenicity. J. exp. Med. *133:* 1171 (1971).
570 Hedley, D. W.; Currie, G. A.: Monocytes and macrophages in malignant melanoma. III. Reduction of nitroblue tetrozolium by peripheral blood monocytes. Br. J. Cancer *37:* 747 (1978).
571 Hefeneider, S. H.; Conlon, P. J.; Henney, C. S.; Gillis, S.: In vivo interleukin 2 administration augments the generation of alloreactive cytolytic T lymphocytes and resident natural killer cells. J. Immun. *130:* 222 (1983).
572 Heijnen, C. J.; Uytdehaag, F.; Pot, K. H.; Ballieux, R. E.: Antigen-specific human T cell factors. I. T cell helper factor: Biologic properties. J. Immun. *126:* 497 (1981).
573 Hellström, I.: A colony inhibition (CI) technique for demonstration of tumor cell destruction by lymphoid cells in vitro. Int. J. Cancer *2:* 65 (1967).
574 Hellström, I.; Sjögren, H. O.: Demonstration of common specific antigen(s) in mouse and hamster polyoma tumors. Int. J. Cancer *1:* 481 (1966).
575 Hellström, I.; Evans, C. A.; Hellström, K. E.: Cellular immunity and its serum-mediated inhibition in shope-virus-induced rabbit papillomas. Int. J. Cancer *4:* 601 (1969).
576 Hellström, I.; Hellström, K. E.: Studies on cellular immunity and its serum-mediated inhibition in Moloney-virus-induced mouse sarcomas. Int. J. Cancer *4:* 587 (1969).
577 Hellström, I.; Hellström, K. E.: Colony inhibition studies on blocking and nonblocking serum effects on cellular immunity to Moloney sarcomas. Int. J. Cancer *5:* 195 (1970).
578 Hellström, I.; Hellström, K. E.; Bill, A. H.; Pierce, G. E.; Yang, J. P. S.: Studies on cellular immunity to human neuroblastoma cells. Int. J. Cancer *6:* 172 (1970).
579 Hellström, I.; Hellström, K. E.; Sjögren, H. O.; Warner, G. A.: Demonstration of cell-mediated immunity to human neoplasms of various histological types. Int. J. Cancer *7:* 1 (1971).
580 Hellström, I.; Sjögren, H. P.; Warner, G. A.: Blocking of cell-mediated tumour immunity by sera from patients with growing neoplasms. Int. J. Cancer *7:* 226 (1971).
581 Hellström, K. E.; Hellström, I.: Lymphocyte mediated cytotoxicity and blocking serum activity to tumor antigen. Adv. Immunol. *18:* 209 (1974).
582 Hellström, K. E.; Hellström, I.: The role of cell-mediated immunity in control and growth of tumors. Clin. Immunobiol. *2:* 233 (1974).
583 Hellström, I.; Hellström, K. E.; van Belle, G.; Warner, G. A.: Leukocyte-mediated reactivity to human tumors as detected by leukocyte adherence inhibition (LAI) test. I. Demonstration of tumor type specific reactions. Am. J. clin. Path. *68:* 706 (1977).

584 Hellström, I.; Brown, J. P.; Klitzman, J. M.; Hellström, K. E.: A highly sensitive, and reproducible microcytotoxicity assay for demonstrating cytotoxic antibodies to cell surface antigens. J. immunol. Methods 22: 369 (1978).
585 Hellström, I.; Brown, J. P.; Hellström, K. E.: Monoclonal antibodies to two determinants of melanoma-antigen p 97 act synergistically in complement-dependent cytotoxicity. J. Immun. 127: 157 (1981).
586 Hellström, I.; Hellström, K. E.; Yeh, M.-Y.: Lymphocyte-dependent antibodies to antigens 3.1, a cell-surface antigen expressed by a subgroup of human melanomas. Int. J. Cancer 27: 281 (1981).
587 Hellström, K. E.; Hellström, I.; Brown, J. P.: Human tumor-associated antigens identified by monoclonal antibodies. Semin. Immunopathol. 5: 127 (1982).
588 Henderson, D. C.; Parker, D.; Turk, J. L.: Dissociation between tumour resistance and delayed-type hypersensitivity to tumour-associated antigens in the mouse. Immunology 39: 1 (1980).
589 Hengst, J. C. D.; Mokyr, M. B.; Dray, S.: Cooperation between cyclophosphamide tumoricidal activity and host antitumor immunity in the cure of mice-bearing large MOPC-315 tumors. Cancer Res. 41: 2163 (1981).
590 Henle, G.; Henle, W.; Klein, G.: Demonstration of two distinct components in the early antigen complex of Epstein-Barr virus-infected cells. Int. J. Cancer 8: 272 (1971).
591 Henle, W.; Ho, H. C.; Kwan, H. C.: Antibodies to Epstein-Barr virus-related antigens in nasopharyngeal carcinoma. Comparison of active cases with long-term survivors. J. natn. Cancer Inst. 51: 361 (1973).
592 Henney, C. S.: Mechanisms of tumor cell destruction; in Green, Cohen, Mechanisms of tumor immunity, p. 55 (John Wiley and Sons, New York 1977).
593 Henney, C. S.: T-cell-mediated cytolysis: An overview of some current issue. Contemp. Top. Immunobiol. 7: 245 (1977).
594 Henney, C. S. et al.: The role of cylic 3′, 5′ adenosine monophosphate in the specific cytolytic activity of lymphocytes. J. Immun. 108: 1526 (1972).
595 Henney, C. S.; Tracey, D.; Durdik, J. M.; Klimpel, G.: Natural killer cells in vitro and in vivo. Am. J. Path. 93: 459 (1978).
596 Henson, P. M.: Interaction of cells with immune complexes: Adherence release of constituents and tissue injury. J. exp. Med. 134: 114 (1971).
597 Henson, P. M.: Complement-dependent platelet and polymorphonuclear leukocyte reactions. Transplant. Proc. 6: 27 (1974).
598 Heppner, G.; Henry, E.; Stolbach, L.; Cummings, F.; McDonough, E.; Calabresi, P.: Problems in the clinical use of the microcytotoxicity assay for measuring cell-mediated immunity to tumor cells. Cancer Res. 35: 1931 (1975).
599 Herberman, R. B.: Immunogenicity of tumor antigens. Biochim. biophys. Acta 473: 93 (1977).
600 Herberman, R. B.: Immunologic tests in diagnosis of cancer. Am. J. clin. Path. 68: 688 (1977).
601 Herberman, R. B.: Immunoregulation and natural killer cells. Mol. Immunol. 19: 1313 (1982).
602 Herberman, R. B.; Nunn, M. E.; Holden, H. T.; Lavrin, D. H.: Natural cytotoxic reactivity of mouse lymphoid cells against syngeneic and allogeneic tumors. II. Characterization of effector cells. Int. J. Cancer 16: 230 (1975).

Literatur 125

603 Herberman, R. B.; Oldham, R. K.: Problems associated with study of cell-mediated immunity of human tumors by microcytotoxicity assays. J. natn. Cancer Inst. 55: 749 (1975).
604 Herberman, R. B.; Holden, H. T.; Ting, C. C.; Lavrin, D. L.; Kirchner, H.: Cell-mediated immunity to leukemia virus- and tumor-associated antigens in mice. Cancer Res. 36: 615 (1976).
605 Herberman, R. B.; Nunn, M. E.; Holden, H. T.; Staal, S.; Djeu, J. Y.: Augmentation of natural cytotoxic reactivity of mouse lymphoid cells against syngeneic and allogeneic target cells. Int. J. Cancer 19: 555 (1977).
606 Herberman, R. B.; Holden, H. T.: Natural cell-mediated immunity. Adv. Cancer Res. 27: 305 (1978).
607 Herberman, R. B.; Nunn, M. E.; Holden, H. T.: Low density of Thy-1 antigen on mouse effector cells mediating natural cytotoxicity against tumor cells. J. Immunol. 121: 304 (1978).
608 Herberman, R. B.; Djeu, J. Y.; Kay, H. D.; Ortaldo, J. R.; Riccardi, C.; Bonnard, G. D.; Holden, H. T.; Fagnani, R.; Santoni, A.; Puccetti, P.: Natural killer cells: Characteristics and regulation of activity. Immunol. Rev. 44: 43 (1979).
609 Herberman, R. R.; Ortaldo, J. R.; Bonnard, G. D.: Augmentation by interferon of human natural and antibody-dependent cell-mediated cytotoxicity. Nature 277: 221 (1979).
610 Herberman, R. B.; Ortaldo, J. R.: Natural killer cells: Their role in defenses against disease. Science 214: 24 (1981).
611 Herbrink, P.; Van Bussel, F. J.; Warnaar, S. O.: The antigen spot test (AST): A highly sensitive assay for the detection of antibodies. J. immunol. Methods 48: 293 (1982).
612 Hericourt, J.; Richet, C.: «Physiologie Pathologique» – de la serotherapie dans le traitement du cancer. C. r. hebd. Séanc. Acad. Sci., Paris 121: 567 (1895).
613 Herlyn, D.; Herlyn, M.; Steplewski, Z.; Koprowski, H.: Monoclonal antibodies in cell-mediated cytotoxicity against human melanoma and colorectal carcinoma. Eur. J. Immunol. 9: 657 (1979).
614 Herlyn, M.; Steplewski, Z.; Herlyn, D.; Koprowski, H.: Colorectal carcinoma-specific antigen: Detection by means of monoclonal antibodies. Proc. natn. Acad. Sci. USA 76: 1438 (1979).
615 Herlyn, D. M.; Steplewski, Z.; Herlyn, M. F.; Koprowski, H.: Inhibition of growth of colorectal carcinoma in nude mice by monoclonal antibody. Cancer Res. 40: 717 (1980).
616 Herlyn, D. M.; Koprowski, H.: Monoclonal anticolon carcinoma antibodies in complement-dependent cytotoxicity. Int. J. Cancer 27: 769 (1981).
617 Heron, I.; Berg, K.; Cantell, K.: Regulatory effect of interferon on T cells in vitro. J. Immun. 117: 1370 (1976).
618 Heron, I.; Hokland, M.; Moller-Larsen, A.; Berg, K.: The effect of interferon on lymphocyte-mediated effector cell functions: Selective enhancement of natural killer cells. Cell. Imunol. 42: 183 (1979).
619 Hersey, P.; Edwards, A.; Edwards, J.; Adams, E.; Milton, G. W.; Nelson, D. S.: Specificity of cell-mediated cytotoxicity against human melanoma lines: Evidence for non-specific killing by activated T cells. Int. J. Cancer 16: 173 (1975).
620 Hersey, P.; Edwards, A.; Milton, G. W.; McCarthy, W. H.: Relationship of cell-

mediated cytotoxicity against melanoma cells to prognosis in melanoma patients. Br. J. Cancer *37:* 505 (1978).
621 Hersey, P.; Edwards, A.; McCarthy, W. H.: Tumour-related changes in natural killer cell activity in melanoma patients. Influence of stage of disease, tumor thickness and age of patients. Int. J. Cancer *25:* 187 (1980).
622 Hersh, E. M.; Quesada, J.; Murphy, S. G.; Gutterman, J. U.; Hutchins, R. D.: Evaluation of therapy with methanol extraction residue of BCG (MER). Cancer Immunol. Immunother. *14:* 4 (1982).
623 Hervé, P.; Philip, T.; Flesch, M.; Rozenbaum, A.; Plouvier, E.; Cahn, J. Y.; Noir, A.; Peters, A.; Leconte des Floris, R.: Intensive cytoreductive regimen and autologous bone marrow transplantation in leukemia. Present status and the future. A review. Eur. J. clin. Oncol. *19:* 1043 (1983).
624 Hewitt, H. B.; Blake, E. R.; Walder, A. S.: A critique of the evidence for active host defence against cancer, based on personal studies of 27 murine tumours of spontaneous origin. Br. J. Cancer *33:* 241 (1976).
625 Hibbs, J. B., Jr.: Macrophage nonimmunologic recognition: Target cell factors related to contact inhibition. Science *180:* 868 (1973).
626 Hibbs, J. B.: Discrimination between neoplastic and non-neoplastic cells in vitro by activated macrophages. J. natn. Cancer Inst. *53:* 1487 (1974).
627 Hibbs, J. B.; Lambert, L. H.; Remington, J. S.: Control of carcinogenesis: a possible role for the activated macrophage. Science *177:* 998 (1972).
628 Hibbs, J. B., Jr.; Lambert, L. H., Jr.; Remington, J. S.: Possible role of macrophage mediated nonspecific cytotoxicity in tumor resistance. Nature, Lond. New Biol. *235:* 48 (1972).
629 Hibbs, J. B.; Taintor, R. R.; Chapman, H. A.; Weinberg, J. B.: Macrophage tumor killing influence of the local environment. Science *197:* 279 (1977).
630 Hibbs, J. B.; Chapman, H. A.; Weinberg, J. B.: The macrophage as an anti-neoplastic surveillance cell: Biological perspectives. J. reticuloendoth. Soc. *24:* 549 (1978).
631 Hibbs, J. B., Jr.; Chapman, H. A., Jr.; Weinberg, J. B.: Regulation of macrophage non-specific tumoricidal capability; in van Furth, Mononuclear phagocytes. Functional aspects, part II, p. 67 (Nijhoff, The Hague 1980).
632 Higgins, T. J.; Parish, C. R.: Extraction of the carbohydrate-defined class of Ia antigens from murine spleen cells and serum. Mol. Immunol. *17:* 1065 (1980).
633 Hilgers, J.; Sonnenberg, A.; Nusse, R.: Antigenic modulation of mammary tumor virus envelope antigen on GR thymic lymphoma cells in relation to expression of H-2, TL cell surface antigens and Thy 1. Br. J. Cancer *42:* 542 (1980).
634 Hinuma, Y.; Nagata, K.; Hanaoka, M.; Nakai, M.; Maisumoto, T.; Kinoshita, T.; Shirakawa, S.; Miyoshi, I.: Adult T-cell leukemia: Antigen in an ATL cell line and detection of antibodies to the antigen in human sera. Proc. natn. Acad. Sci. USA *78:* 6476 (1981).
635 Hinuma, Y.; Komoda, H.; Chosa, T.; Kondo, T.; Kohakura, M.; Takenaka, T.; Kikuchi, M.; Ichimaru, M.; Yunoki, K.; Sato, I.; Matsuo, R.; Takiuchi, Y.; Uchino, H.; Hanaoka, M.: Antibodies to adult T-cell leukemia-virus-associated antigen (ATLA) in sera from patients with ATL and controls in Japan: A nation-wide seroepidemiologic study. Int. J. Cancer *29:* 631 (1982).
636 Hirsch, M. S.; Ellis, D. A.; Black, R. H.: Leukemia virus activation during homograft rejection. Science *180:* 55 (1973).

637 Hirschberg, H.; Bergh, O. J.; Thorsby, E.: Antigen presenting properties of human vascular endothelial cells. J. exp. Med. *152:* 2495 (1980).

638 Hirschberg, H.; Braathen, L. R.; Thorsby, E.: Antigen presentation by vascular endothelial cells and epidermal Langerhans cells: The role of HLA-DR. Immunol. Rev. *66:* 57 (1982).

639 Hodes, R. J.; Hathcock, K. S.: In vitro generation of suppressor cell activity: suppression of in vitro induction of cell-mediated cytotoxicity. J. Immunol. *116:* 167 (1976).

640 Höffken, K.; Schmidt, C. G.: Tumorantigene. Dt. med. Wschr. *103:* 1187 (1978).

641 Hoffmann, M. K.; Kappler, J. W.: Two distinct mechanisms of immune suppression by antibody. Nature *272:* 64 (1978).

642 Hofstaetter, T.; Gronski, P.; Seiler, F. R.: Immunotoxins – Theoretical and practical aspects. Behring Inst. Mitt. *74:* 113 (1984).

643 Holan, V.; Hasek, M.; Bubenick, J.; Chutna, J.: Antigen-mediated macrophage adherence inhibition. Cell. Immunol. *13:* 107 (1974).

644 Holden, H. T.; Landolfo, S.; Herberman, R. B.: Cell-dependent reactivity against tumor-associated antigens on allogeneic target cells. Transplant. Proc. *9:* 1149 (1977).

645 Holdener, E. E.; Schnell, P.; Spieler, P.; Senn, H.: In vitro effect of Interferon-α on human granulocyte/macrophage progenitor cells and human clonogenic tumor cells. Cancer Res. *94:* 205 (1984).

646 Holland, J. R.: Ex pluribus unum: Presidential address to American Association for Cancer Research. Cancer Res. *31:* 1319 (1971).

647 Holland, J. R.; St. Arneault, G.; Bekesi, G.: Combined chemo- and immunotherapy of transplantable and spontaneous murine leukemia. Proc. Am. Ass. Cancer Res. *13:* 83 (1972).

648 Holland, J. F.; Bekesi, J. G.: Immunotherapy of human leukemia with neuraminidase-modified cells. Med. Clins N. Am. *60:* 539 (1976).

649 Holland, J. F.; Bekesi, J. G.; Cuttner, J.; Glidewell, O.: Chemoimmunotherapy in acute myelocytic leukemia. Israel J. med. Scis *13:* 694 (1977).

650 Holland, J. F.; Bekesi, J. G.: Comparison of chemotherapy with chemotherapy plus VCN-treated cells in acute myelocytic leukemia; in Terry, Windhorst, Immunotherapy of cancer: Present status of trials in man, p. 347 (Raven Press, New York 1979).

651 Hollinshead, A. C.; Stewart, T. H. M.; Herberman, R. B.: Delayed-hypersensitivity reactions to soluble membrane antigens of human malignant lung cells. J. natn. Cancer Inst. *52:* 327 (1974).

652 Hollinshead, A. C.; Stewart, T. H. M.: Specific and nonspecific immunotherapy as an adjunct to curative surgery for cancer of the lung. Yale J. Biol. Med. *54:* 367 (1981).

653 Hollinshead, A.; Elias, E. G.; Arlen, M.; Buda, B.; Mosley, M.; Scherrer, J.: Specific active immunotherapy in patients with adenocarcinoma of the colon utilizing tumor-associated antigens (TAA). Cancer *56:* 480 (1985).

654 Holmes, E. C.; Kohon, B. D.; Morton, D. L.: Soluble tumor-specific transplantation antigens from methylcholanthrene-induced guinea pig sarcomas. Cancer *25:* 373 (1970).

655 Holtkamp, B.; Fisher-Lindahl, K.; Segall, M.; Rajewsky, K.: Spontaneous loss and subsequent stimulation of H-2 expression in clones of a heterozygous lymphoma cell line. Immunogen. *9:* 405 (1979).

656 Holzmann, B.; Johnson, J. P.; Kaudewitz, P.; Riethmüller, G.: In situ analysis of antigens on malignant and benign cells of the melanocyte lineage. J. exp. Med. *161:* 366 (1985).

657 Hoover, H. C.; Surdyke, M. G.; Dangel, R. B.; Peters, L. C.; Hanna, M. G.: Prospectively randomized trial of adjuvant active-specific immunotherapy for human colorectal cancer. Cancer *55:* 1236 (1985).

658 Hopkins, N.; Besmer, P.; de Leo, A. B.; Law, L. W.: High frequency co-transfer of the transformed phenotype and a TSTA using DNA from the 3-methylcholanthrene-induced Meth A sarcoma of Balb/c mice. Proc. natn. Acad. Sci. USA *78:* 7555 (1981).

659 Hopper, D. G.; Pimm, M. V.; Baldwin, R. W.: Silica abrogation of mycobacterial adjuvant contact suppression of tumour growth in rats and athymic mice. Cancer Immunol. Immunother. *1:* 143 (1976).

660 Hopper, K. E.; Harrison, J.; Nelson, D. S.: Partial characterization of anti-tumor effector macrophages in the peritoneal cavities of concomitantly immune mice and mice injected with macrophage-stimulating agents. J. reticuloendoth. Soc. *26:* 259 (1979).

661 Hopper, K. E.; Wood, P. R.; Nelson, D. S.: Macrophage heterogeneity. Vox Sang. *36:* 257 (1979).

662 Hosoi, S.; Nakamura, T.; Higashi, S.; Yamamuro, T.; Toyama, S.; Shinomiya, K.; Mikawa, H.: Detection of human osteosarcoma-associated antigen(s) by monoclonal antibodies. Cancer Res. *42:* 654 (1982).

663 Houghton, A. N.; Mintzer, D.; Cordon-Cardo, C.; Welt, S.; Fliegel, B.; Vadhan, S.; Carswell, E.; Melamed, M. R.; Oettgen, H. F.; Old, L. J.: Mouse monoclonal IgG3 antibody detecting G_{D3} ganglioside: A phase I trial in patients with malignant melanoma. Proc. natn. Acad. Sci. USA *82:* 1242 (1985).

664 Howard, D. R.; Taylor, C. R.: An antitumor antibody in normal human serum. Oncology *37:* 142 (1980).

665 Howell, S. B.; Dean, J. H.; Law, L. W.: Defects in cell-mediated immunity during growth of a syngeneic simian virus-induced tumor. Int. J. Cancer *15:* 152 (1975).

666 Huang, K. Y.; Donahoe, R. M.; Gordon, F.; Dressler, H. R.: Enhancement of phagocytosis by interferon-containing preparations. Infect. Immunity *4:* 581 (1971).

667 Huang, L. C.; Brockhaus, M.; Magnani, J. L.; Cuttitta, F.; Rosen, S.; Minna, J. D.; Ginsburg, V.: Many monoclonal antibodies with an apparent specificity for certain lung cancers are directed against a sugar sequence found in lacto-N-fucopentaose III. Archs Biochem. Biophys. *220:* 318 (1983).

668 Hudig, D.; Redelman, D.; Mendelsohn, J.: Inhibition of human natural cytotoxicity by proteinase substrates (Abstract 469). Fed. Proc. *39:* 359 (1980).

669 Huebner, R. J.; Rowe, W. P.; Turner, H. C.; Lane, W. T.: Specific adenovirus complement-fixing antigens in virus-free hamster and rat tumours. Proc. natn. Acad. Sci. USA *50:* 379 (1963).

670 Hughes, L. F.; Kearney, R.; Tully, M.: A study in clinical cancer immunotherapy. Cancer *26:* 269 (1970).

671 Hulliger, K.; Blazkovec, A. A.; Sorkin, E.: A study of the passive cellular transfer of local cutaneous hypersensitivity. IV. Transfer of hypersensitivity to sheep erythrocytes with peritoneal exudate cells coated with antibody. Int. Arch. Allergy *33:* 281 (1968).

672 Humes, J. L.; Bonney, R. J.; Pelus, L.; Dahlgren, M. E.; Sadowski, S. J.; Kuehl, F. A., Jr.; Davies, P.: Macrophages synthesis and release prostaglandins in response to inflammatory stimuli. Nature *269:* 149 (1977).

673 Humphrey, J. H.: Differentiation of function among macrophages. In: Microenvironments in haemopoietic and lymphoid differentiation. Ciba Foundation Symp. *84:* 302 (1981).

674 Humphrey, J. L.; Jewell, W. R.; Murray, O. R.; Griffen, W. O.: Immunotherapy for the patient with cancer. Ann. Surg. *173:* 47 (1971).

675 Humphrey, L. J.; Murray, D. R.; Boehm, O. R.; Jewell, W. R.; Griffen, W. O., Jr.: Immunotherapy of cancer in man. VII. Studies of patients with cancer of the alimentary tract. Am. J. Surg. *121:* 165 (1971).

676 Hunter-Craig, I.; Newton, K. A.; Westbury, G.; Lacey, B. W.: Use of vaccinia virus in the treatment of metastatic malignant melanoma. Br. med. J. *ii:* 512 (1970).

677 Hutchin, P.; Amos, D. B.; Prioleau, W. H., Jr.: Interactions of humoral antibodies and immune lymphocytes. Transplantation *5:* 68 (1967).

678 Hutchinson, I. V.: Antigen-reactive cell opsonisation and its role in antibody-mediated immune suppression. Immunol. Rev. *49:* 167 (1980).

679 Hutchinson, I. V.; Roman, J.; Bonavida, B.: Opsonisation of anti-tumor reactive lymphocytes in SJL/J mice-bearing spontaneous or transplanted reticulum cell sarcomas. Adv. exp. med. Biol. *121 B:* 553 (1980).

680 Ignarro, L. J.: Regulation of lysosomal enzyme secretion: Role in inflammation. Agents and Actions *4:* 241 (1974).

681 Imai, K.; Ng, A.-K.; Ferrone, S.: Characterization of monoclonal antibodies to human melanoma-associated antigens. J. natn. Cancer Inst. *66:* 489 (1981).

682 Imai, K.; Ng, A. K.; Glassy, M. C.; Ferrone, S.: ADCC of cultured human melanoma cells: Analysis with monoclonal antibodies to human melanoma-associated antigens. Scand. J. Immunol. *14:* 369 (1981).

683 Imai, K.; Ng, A. K.; Glassy, M. C.; Ferrone, S.: Differential effect of interferon on the expression of tumor-associated antigens and histocompatibility antigens on human melanoma cells: relationship to susceptibility to immune lysis mediated by monoclonal antibodies. J. Immun. *127:* 505 (1981).

684 Imai, K.; Pellegrino, M. A.; Wilson, B. S.; Ferrone, S.: Higher cytolytic efficiency of an IgG2 alpha than of an IgG1 monoclonal antibody reacting with the same (or spatially close) determinant on a human high-molecular-weight melanoma-associated antigen. Cell. Immunol. *72:* 239 (1982).

685 Imanishi, J.; Yokota, Y.; Kishida, T.; Mukainaka, T.; Matsuo, A.: Phagocytosis-enhancing effect of human leukocyte interferon, preparations of human peripheral monocytes in vitro. Acta Virol. *19:* 52 (1975).

686 Imanishi, J.; Oishi, K.; Kishida, T.; Negoro, Y.; Izuka, M.: Effects of interferon preparations on rabbit corneal xenograft. Archs Virol. *53:* 157 (1977).

687 Invernizzi, G.; Parmiani, G.: Tumor-associated transplantation antigens of chemically induced sarcomata cross-reacting with allogeneic histocompatibility antigens. Nature *254:*713 (1975).

688 Ioachim, H. L.: The stromal reaction of tumors: An expression of immune surveillance. J. natn. Cancer Inst. *57:* 465 (1976).

689 Ioachim, H. L.; Keller, S.; Sabbath, M.: Antigenic expression as a determining factor

of tumor growth in Gross virus lymphoma; in Richards, Immunology of cancer. Progress in experimental tumor research, vol. 19, p. 284 (Karger, Basel 1974).

690 Irie, K.; Irie, R. F.; Morton, D. L.: Evidence for in vivo reaction of antibody and complement to surface antigens of human cancer cells. Science 185: 454 (1974).

691 Irvin, G. L.; Eustace, J. C.: The enhancement and rejection of tumor allografts by immune lymph node cells. Transplantation 10: 555 (1970).

692 Irvin, G. L.; Eustace, J. C.: A study of tumor allograft-sensitized lymph nodes in mice. I. Biologic activities of transferred cells and antibody titers of donor and recipient mice. J. Immun. 106: 956 (1971).

693 Isaacs, A.; Lindemann, I.: Virus interferons. I. The interferon. Proc. R. Soc. 147:268 (1957).

694 Ishidate, M.; Hashimoto, Y.; Odashima, S.; Sudo, H.: Studies on acquired transplantation resistance. I. Pretreatment of donryu rat with attenuated yoshida sarcoma cells. Gann 56: 13 (1965).

695 Ishizaka, K.; Adachi, T.: Generation of specific helper cells and suppressor cells in vitro for the IgE and IgG antibody responses. J. Immunol. 117: 40 (1976).

696 Ishizuka, M.; Aoyagi, T.; Takeuchi, T.; Umezawa, H.: Activity of bestatin: Enhancement of immune responses and antitumor effect; in Umezawa, Small molecular immunomodifiers of microbial origin: Fundamental and clinical studies of Bestatin, p. 17 (Pergamon Press, New York 1981).

697 Israel, L. et al.: Antitumor activity of tilorone in man: A preliminary report; in Chirigos, Modulation of host immune resistance in the prevention or treatment of induced neoplasias, p. 145 (DHEW Publication No. (NIH), 1974).

698 Iyer, G. Y. N.; Islam, M. F.; Qustel, J. H.: Biochemical aspects of phagocytosis. Nature 192: 535 (1961).

699 James, K.; Cullen, R. T.; Milne, I.; Norval, M.: Anti-tumour response induced by short-term pretreatment with tumour cells. Br. J. Cancer 37: 269 (1978).

700 Jandinsky, J. J.; Li, J.; Wettstein, P. J.; Frelinger, J. A.; Scott, D. W.: Role of self carriers in the immune response and tolerance. V. Reversal of trinitrophenyl-modified self suppression of the B-cell response by blocking of H-2 antigens. J. exp. Med. 151: 133 (1980).

701 Jasmin, C.; Piton, C.; Rosenfeld, C.: Effets de l'Iodoacétamide sur les Cellules de la Leucémie Virale de Rauscher. Int. J. Cancer 3: 254 (1968).

702 Jeejeebhoy, H. F.: Stimulation of tumor growth by the immune response. Int. J. Cancer 13: 665 (1974).

703 Jerne, N. K.: The somatic generation of immune recognition. Eur. J. Immunol. 1: 1 (1971).

704 Jerne, N. K.: Towards a network theory of the immune system. Ann. Immunol., Paris 125 C: 373 (1974).

705 Jett, J. R.; Mantovani, A.; Herberman, R. B.: Augmentation of human monocyte-mediated cytolysis by interferon. Cell. Immunol. 54: 425 (1980).

706 Johannsen, R.; Sedlecek, H. H.: Specificity of cytotoxic antibodies to autologous human lymphocytes treated with neuraminidase from Vibrio cholerae. Behring Inst. Mitt. 55: 209 (1974).

707 Johannsen, R.; Carlsson, A. B.; Sedlacek, H. H.: In vitro transformation of human lymphocytes by neuraminidase from Vibrio cholerae (VCN) (Abstracts, p. 86) 6th Workshop on Leukocyte Cultures, Basel, March 17–19 (1975).

708 Johannsen, R.; Haupt, H.; Bohn, H.; Heide, K.; Seiler, F. R.; Schwick, H. G.: Inhibition of the mixed leukocyte culture (MLC) by proteins: Mechanism and specificity of the reaction. Z. ImmunForsch. *152:* 280 (1976).

709 Johannsen, R.; Sedlacek, H. H.; Seiler, F. R.: Adjuvant effect of Vibrio cholerae neuraminidase on the in vitro and in vivo immune response; in Rainer, Proc. of the Symp. «Immunotherapy of Malignant Disease», p. 244 (Schattauer, Stuttgart 1978).

710 Johannsen, R.; Sedlacek, H. H.; Schmidtberger, R.; Schick, H. J.; Seiler, F. R.: Characteristics of cytotoxic antibodies against neuraminidase-treated lymphocytes in man. J. natn. Cancer Inst. *62:* 733 (1979).

711 Johannsen, R.; Sedlacek, H. H.; Seiler, F. R.: The neuraminidase-induced enhanced immune response can be attributed to immune stimulatory and antigenic properties of the enzyme. Transplant. Proc. *11:* 1411 (1979).

712 Johnson, G. D.; Holborrow, E. J.: Immunofluorescence; in Weir, Handbook experimental immunology; 2nd ed., p. 18.1 (Blackwell Scientific Public., Oxford, London 1973).

713 Johnson, H. N.; Smith, B. G.; Baron, S.: Inhibition of the primary in vitro antibody response by interferon preparations. J. Immunol. *114:* 403 (1975).

714 Johnson, H. M.; Baron, S.: The nature of the suppressive effect of interferon and interferon inducers on the in vitro immune response. Cell. Immunol. *25:* 106 (1976).

715 Johnson, M. H.; Walker, R. W. H.; Keir, G.; Thompson, E. J.: A new method for identification of proteins separated in poly-acrylamide gels. Biochem. Soc. Transact. *10:* 32 (1980).

716 Johnson, J. P.; Demmer-Dieckmann, M.; Meo, T.; Hadam, M. R.; Riethmüller, G.: Surface antigens of human melanoma cells defined by monoclonal antibodies. I. Biochemical characterization of two antigens found on cell lines and fresh tumors of diverse tissue origin. Eur. J. Immunol. *11:* 825 (1981).

717 Johnston, R. B., Jr.; Chadwick, D. A.; Pabst, M. J.: Release of superoxide anion by macrophages: Effect of in vivo or in vitro priming; in van Furth, Mononuclear phagocytes: Functional aspects, part II, p. 44 (Nijhoff, The Hague 1980).

718 Jolles, P.; Paraf, A.: Mechanisms of adjuvant activity; in Kleinzeller et al., Chemical and biological basis of adjuvants: Molecular biology, biochemistry and biophysics, vol. 13, p. 81 (Springer, Berlin 1973).

719 Jolley, G. M.; Boyle, M. D. P.; Ormerod, M. G.: The destruction of allogeneic tumor cells by antibody and adherent cells from peritoneal cavities of mice. Cell. Immunol. *22:* 262 (1976).

720 Jondal, M.; Holm, G.; Wigzell, H.: Surface markers on human T and B lymphocytes. I. A large population of lymphocytes forming non-immune rosettes with SRBC. J. exp. Med. *136:* 207 (1972).

721 Jondal, M.; Klein, G.: Classification of lymphocytes in nasopharyngeal carcinoma (NPC) biopsies. Biomedicine *23:* 163 (1975).

722 Jondal, M.; Svedmyr, E.; Klein, E.; Singh, S.: Killer T cells in a Burkitt's lymphoma biopsy. Nature *266:* 405 (1975).

723 Jones, P. D.; Castro, J. E.: Immunological mechanism in metastatic spread and the antimetastatic effect of C. parvum. Br. J. Cancer *35:* 519 (1977).

724 Jose, D. G.; Seshadri, R.: Circulating immune complexes in human neuroblastoma:

Direct assay and role in blocking specific cellular immunity. Int. J. Cancer *13:* 824 (1974).
725 Juillard, G. J. F.; Boyer, P. J. J.; Yamashiro, C. H.: A phase I study of active specific intralymphatic immunotherapy (ASILI). Cancer *41:* 2215 (1978).
726 Kabat, E. A.; Bezer, A. E.: The effect of variation in molecular weight on the antigenicity of dextran in man. Archs Biochem. Biophys. *78:* 306 (1958).
727 Kadish, A. S.; Marcus, D. M.; Bloom, B. R.: Inhibition of leukocyte migration by human breast cancer-associated antigens. Int. J. Cancer *18:* 581 (1976).
728 Kahan, B. D.; Holmes, E. C.; Reisfeld, R. A.; Morton. D. L.: Water soluble guinea pig transplantation antigen from carcinogen-induced sarcomas. J. Immun. *102:* 28 (1969).
729 Kahan, B. D. et al.: Immunotherapeutic effects of tumor-specific transplantation antigens released by 1-Blutanol. Cancer *49:* 1168 (1982).
730 Kalina, M.; Berke, G.: Contact regions of cytotoxic T lymphocate-target cell conjugates. Cell. Immunol. *25:* 41 (1976).
731 Kaliss, N.: Immunological enhancement of tumor homografts in mice: A review. Cancer Res. *18:* 992 (1958).
732 Kamo, I.; Patel, C.; Kateley, J.: Immunosuppression induced in vitro by mastocytoma tumor cells and cell-free extracts. J. Immun. *114:* 1749 (1975).
733 Kamoun, M.; Martin, P. J.; Hansen, J. A.; Brown, M. A.; Siadek, A. N.; Nowinski, R. C.: Identification of a human T lymphocyte surface protein associated with the E-rosette receptor. J. exp. Med. *153:* 207 (1981).
734 Kampschmidt, R. F.; Upchurch, H. F.: Stimulation of the reticuloendothelial system in tumor-bearing rats. J. reticuloendoth. Soc. *5:* 510 (1968).
735 Kampschmidt, R. F.; Pulliam, L. A.: Changes in the opsonin and cellular influences on phagocytosis during the growth of transplantable tumors. J. reticuloendoth. Soc. *11:* 1 (1972).
736 Kaplan, A. M.; Morahan, P. S.; Regelson, W.: Induction of macrophage-mediated tumor-cell cytotoxicity by pyran copolymer. J. natn. Cancer Inst. *52:* 1919 (1974).
737 Kaplan, G.; Olstad, R.: Heterogeneity in surface glycoproteins of mouse peritoneal macrophage populations. Expl Cell Res. *135:* 379 (1981).
738 Kapp, J. A.; Pierce, C. W.; Schlossman, B.; Benacerraf, B.: Genetic control of immune responses in vitro. V. Stimulation of suppressor T cells in nonresponder mice by the terpolymer L-glutamic acid**60-L-alanine**30-L-tyrosine**10 (GAT). J. exp. Med. *140:* 648 (1974).
739 Karre, K.; Klein, G. O.; Kiessling, R.; Klein, G.; Roder, J. C.: Low natural in vivo resistance to syngeneic leukemias in natural killer-deficient mice. Nature, Lond. *284:* 624 (1980).
740 Kasai, M.; Leclerc, J. C.; Shen, F. W.; Cantor, H.: Identification of Ly 5 on the surface of natural killer cells in normal and athymic inbred mouse strains. Immunogen. *8:* 153 (1979).
741 Kasai, M.; Saxton, R. E.; Holmes, E. C.; Burk, M. W.; Morton, D. L.: Membrane antigens detected on human lung carcinoma cells by hybridoma monoclonal antibody. J. surg. Res. *30:* 403 (1981).
742 Kassel, R. L.; Old, L. J.; Carswell, E. A.; Fiore, N. C.; Hardy, W. D., Jr.: Serummediated leukemia cell destruction in AKR mice: Role of complement in the phenomenon. J. exp. Med. *138:* 925 (1973).

743 Kataoka, T.; Oh-Hashi, F.; Tsukagoshi, S.; Sakurai, Y.: Induction of resistance to L1210 leukemia in Balb/c × DBA2CrF$_1$ mice, with L1210 cells treated with glutaraldehyde and concanavalin A. Cancer Res. 37: 964 (1977).

744 Katoh, T.; Bosslet, K.; Kurrle, R.; Sedlacek, H.-H.; Seiler, F. R.: Specificity of murine monoclonal antibodies induced by a choriocarcinoma cell-line (BEWO). Behring Inst. Mitt. 74: 49 (1984).

745 Katz, D. H.; Benacerraf, B.: The regulatory influence of activated T cells on B cell responses to antigen. Adv. Immunol. 15: 1 (1972).

746 Katz, D. H.; Hamaoka, T.; Dorf, M. E.; Benacerraf, B.: Cell interactions between histoincompatible T and B lymphocytes. The H-2 gene complex determines successful physiologic lymphocyte interactions. Proc. natn. Acad. Sci. USA 70: 2624 (1973).

747 Katzav, S.; De Baetselier, P.; Gorelik, E.; Feldman, M.; Segal, S.: Immunogenetic control of metastasis formation by methylcholanthrene-induced tumor (T10) in mice: Differential expression of H-2 gene products. Transplant. Proc. 8: 742 (1981).

748 Katzav, S.; De Baetselier, P.; Tartakovsky, B.; Feldman, M.; Segal, S.: Alterations in major histocompatibility complex phenotypes of mouse cloned T10 sarcoma cells: Association with shifts from nonmetastatic to metastatic cells. J. natn. Cancer Inst. 71: 317 (1983).

749 Kay, H. D.; Bonnard, W. H.; West, W. H.; Herberman, R. B.: A functional comparison of Fc-receptor-bearing lymphocytes active in natural cytotoxicity and antibody-dependent cellular cytotoxicity. J. Immun. 118: 2054 (1977).

750 Kay, H. D.; Horwitz, D. A.: Evidence by reactivity with hybridoma antibodies for a probable myeloid origin of peripheral blood cells active in natural cytotoxicity and antibody-dependent cell-mediated cytotoxicity. J. clin. Invest. 66: 847 (1980).

751 Kay, M. M. B.; Makinodan, T.: Relationship between aging and the immune system; in Waksman, Ontogeny of the immune system. Progress in allergy, vol. 29, p. 134 (Karger, Basel 1981).

752 Kedar, E.; de Landazuri, M. O.; Fahey, J. L.: Enzymatic enhancement of cell-mediated cytotoxicity and antibody-dependent cell cytotoxicity. J. Immun. 112: 26 (1974).

753 Kedar, E.; Schwartzbach, M.; Raanan, Z.; Hefetz, S.: In vitro induction of cell-mediated immunity to murine leukemia cells. II. Cytotoxic activity in vitro and tumor-neutralizing capacity in vivo of anti-leukemia cytotoxic lymphocytes generated in macrocultures. J. immunol. Methods 16: 39 (1977).

754 Keller, R.: Promotion of tumor growth by antimacrophage agents. J. natn. Cancer Inst. 57: 1355 (1976).

755 Keller, R.: Suppression of natural antitumor defense mechanisms by phorbol esters. Nature, Lond. 282: 729 (1979).

756 Keller, R.: Distinctive characteristics of host tumor resistance in a rat fibrosarcoma model system; in van Furth, Mononuclear phagocytes: Functional aspects part II, p. 68 (Nijhoff, The Hague 1980).

757 Kellock, T. H.; Chambers, H.; Russ, S.: An attempt to procure immunity to malignant disease in man. Lancet i: 217 (1922).

758 Kelly, B. S.; Stredulinsky, U.; Vanden Hoek, J.; Levy, J. G.: Antibodies in the sera of patients with bronchogenic carcinoma that react with antigen from a tumour cell line. Cancer Immunol. Immunother. 12: 5 (1981).

759 Kemp, A. S.; Berke, G.: Inhibition of lymphocyte-mediated cytolysis by the local anaesthetics benzyl and salicyl alcohol. Eur. J. Immunol. *3:* 674 (1973).
760 Kemshead, T.; Walsh, F.; Pritchard, J.: Monoclonal antibody to ganglioside GQ discriminates between haemopoietic cells and infiltrating neuroblastoma tumor cells in bone marrow. Int. J. Cancer *27:* 447 (1981).
761 Kennedy, C. T. C.; Cater, D. B.; Hartveit, F.: Protection of C3H mice against BP-8 tumor by RNA extracted from lymph nodes and spleens of specifically sensitized mice. Acta path. microbiol. scand. *77:* 196 (1969).
762 Kennet, R. H.; Gilbert, F.: Hybrid myelomas producing antibodies against a human neuroblastoma antigen present on fetal brain. Science *203:* 1120 (1979).
763 Kerbel, R. S.; Pross, H. F.: Fc receptor-bearing cells as a reliable marker for quantitation of host lymphoreticular infiltration of progressively growing solid tumors. Int. J. Cancer *18:* 432 (1976).
764 Kerbel, R. S.; Lagarde, E. A.; Dennis, J. W.; Donaghue, T. P.: Spontaneous fusion in vivo between normal cells and tumor cells: possible contribution to tumor progression and metastasis studied with a lectin resistant mutant tumor. Mol. cell. Biol. *3:* 523 (1983).
765 Kern, D. H.; Drogemüller, C. R.; Pilch, Y. H.: Immune cytolysis of rat tumor cells mediated by syngeneic «immune» RNA. J. natn. Cancer Inst. *52:* 299 (1974).
766 Kern, D. H.; Drogemüller, C. R.; Chow, N.; Holleman, D. D.; Pilch, Y. H.: Specificity of anti-tumor immune reactions mediated by xenogeneic immune RNA. J. natn. Cancer Inst. *58:* 117 (1977).
767 Kerney, S. E.; Montague, P. M.; Chretien, P. B.; Nocholson, J. M.; Ekel, T. M.; Hearing, V. J.: Intracellular localization of tumor-associated antigens in murine and human malignant melanoma. Cancer Res. *37:* 1519 (1977).
768 Kersey, J. H.: Lymphoid progenitor cells and acute lymphoblastic leukemia: Studies with monoclonal antibodies. J. clin. Immunol. *1:* 201 (1981).
769 Kersey, J. H.; Spector, B. D.; Good, R. A.: Immunodeficiency and cancer. Adv. Cancer Res. *18:* 211 (1973).
770 Khera, K. S.; Ashkenazi, A.; Rapp, F.; Melnick, J. L.: Immunity in hamsters to cells transformed in vitro and in vivo by SV40. Tests for antigenic relationship among the papovaviruses. J. Immunol. *91:* 604 (1963).
771 Kidd, J. G.: Suppression of growth of Brown-Pearce tumor cells by a specific antibody, with consideration of the nature of reacting cell constituent. J. exp. Med. *83:* 227 (1946).
772 Kiessling, R.; Klein, E.; Wigzell, H.: Natural killer cells in the mouse. I. Cytotoxic cells with specificity for mouse maloney leukemia specificity and distribution according to genotype. Eur. J. Immunol. *5:* 112 (1975).
773 Kiessling, R.; Petrányi, G.; Klein, G.; Wigzell, H.: Genetic variation of in vitro cytolytic activity and in vivo rejection potential of non-immunized semisyngeneic mice against a mouse lymphoma line. Int. J. Cancer *15:* 933 (1975).
774 Kiessling, R.; Petrányi, G.; Klein, G.; Wigzell, H.: Non-T-cell resistance against a mouse Moloney lymphoma. Int. J. Cancer *17:* 275 (1976).
775 Kiessling, R.; Wigzell, H.: An analysis of the murine NK cell as to structure, function and biological relevance. Immunol. Rev. *44:* 165 (1979).
776 Killion, G. J.: The immunotherapeutic value of a L1210 tumour-cell vaccine

depends upon the expression of cell-surface carbohydrates. Cancer Immunol. Immunother. *3:* 87 (1977).
777 Killion, J. J.: Immunotherapy with tumor cell subpopulations. III. Interaction between specific and nonspecific immunostimulants. Cancer Immunol. Immunother. *5:* 27 (1978).
778 Kim, U.: Metastasizing mammary carcinomas in rats: Induction and study of their immunogenicity. Science *167:* 72 (1970).
779 Kim, Z.; Uhlenbruck, G.: Untersuchungen über T-Antigen und T-Agglutinin. Z. ImmunForsch. *130:* 88 (1966).
780 Kim, U.; Baumler, A.; Carruthers, C.; Bielat, K.: Immunological escape mechanism in spontaneously metastasizing mammary tumors. Proc. natn. Acad. Sci. USA *72:* 1012 (1975).
781 Kim, Y. B.; Huh, N. D.; Koren, H. S.; Amos, D. B.: Natural killing (NK) and antibody-dependent cellular cytotoxicity (ADCC) in specific pathogen-free (SPF) miniature swine and germ-free piglets. I. Comparison of NK and ADCC. J. Immun. *125:* 755 (1980).
782 King, G. W.; Lobuglio, A. E.; Lugone, A. L.: Human monocyte glucose metabolism in lymphoma. J. Lab. clin. Med. *89:* 316 (1977).
783 Kirchner, H.: Interferons as antitumor agents. J. Cancer Res. clin. Oncol. *103:* 1 (1982).
784 Kirchner, H.; Chused, T. M.; Herberman, R. B.; Holden, H. T.; Lavrin, D. H.: Evidence of suppressor cell activity in spleens of mice-bearing primary tumors induced by Moloney sarcoma virus. J. exp. Med. *139:* 1473 (1974).
785 Kirkwood, J. M.; Gershon, R. K.: A role for suppressor T cells in immunological enhancement of tumor growth; in Richards, Immunology of cancer. Progress in experimental tumor research, vol. 19, p. 157 (Karger, Basel 1974).
786 Kishida, T.; Morikawa, K.; Ito, H.; Yokota, Y.: Influence de l'interferon sur l'inhibition par les macrophages de la multiplication in vitro de la cellule murine maligne (FM_3A). Comptes rendus de la Societe de Biologie, Paris *167:* 1502 (1973).
787 Kishimoto, T.; Ishizaka, K.: Regulation of antibody response in vitro. VI. Carrier-specific helper cells for IgG and IgE antibody response. J. Immun. *111:* 720 (1973).
788 Klaus, G. G. B.; Humphrey, J. H.; Kunkl, A.; Dungworth, D. W.: The follicular dendritic cell: Its role in antigen presentation in the generation of immunological memory. Immunol. Rev. *53:* 3 (1980).
789 Klavins, J. W.: Advances in biological markers for cancer. Ann. clin. Lab. Sci. *13:* 275 (1983).
790 Klebanoff, S. J.: Oxygen intermediates and the microbicidal event; in van Furth, Mononuclear phagocytes: Functional aspects, part II, p. 43 (Nijhoff, The Hague 1980).
791 Klein, E.: Local cytostatic chemotherapy and immunotherapy. Geriatrics *23:* 154 (1968).
792 Klein, E.: Tumours of skin. 8. Local chemotherapy of metastatic neoplasms. N. Y. St. J. Med. *68:* 900 (1968).
793 Klein, E.: Hypersensitivity reactions at tumor sites. Cancer Res. *29:* 2351 (1969).
794 Klein, P. A.: Adaptation of influenza virus to growth in cultured murine methylcholanthrene induced tumours. Arch. ges. Virusforsch. *45:* 199 (1974).
795 Klein, G.; Sjörgren, H. O.; Klein, E.; Hellström, K. E.: Demonstration of resistance

against methylcholanthrene-induced sarcomas in the primary authochthonous host. Cancer Res. 20: 1561 (1960).
796 Klein, G.; Klein, E.: Antigenic properties of other experimental tumors. Symp. Quant. Biol. 27: 463 (1962).
797 Klein, E.; Holtermann O. A.; Case, R. W.; Milgrom, H.; Rosner, D.; Adler, S.: Responses of neoplasms to local immunotherapy. Am. J. clin. Path. 62: 281 (1974).
798 Klein, G.; Klein, E.: Are methylcholanthrene-induced sarcoma-associated, rejection-inducing (TSTA) antigens modified forms of H-2 or linked determinants? Int. J. Cancer 15: 879 (1975).
799 Klein, G.; Klein, E.: Immune surveillance against virus-induced tumors and non-rejectability of spontaneous tumors: contrasting consequence of host versus tumor evolution. Proc. natn. Acad. Sci. USA 74: 2121 (1977).
800 Klein, G.; Klein, E.: Rejectability of virus-induced tumors and non-rejectability of spontaneous tumors: A lesson in contrasts. Transplant. Proc. 9: 1095 (1977).
801 Klein, M.; Neauport-Santes, C.; Ellerson, J. R.; Fridman, W. H.: Binding site of human IgG subclasses and their domains for Fc-receptors of activated murine T-cells. J. Immun. 119: 1077 (1977).
802 Klein, P. J.; Newman, R. A.; Müller, P.; Uhlenbruck, G.; Citoler, P.; Schaefer, H. E.; Lennartz, K. J.; Fischer, R.: The presence and significance of the Thomsen-Friedenreich antigen in breast tissue. Ii. Its topochemistry in normal, hyperplastic and carcinoma tissue of the breast. Cancer Res. clin. Oncol. 93: 205 (1979).
803 Klein, E.; Case, R. W.; Holtermann, O.; Milgrom, H.; Hahn, G.; Preffer, F.: Clinical effects of local lymphokine administration on neoplastic lesions. Proc. Am. Ass. Cancer Res. 22: 164 (1981).
804 Klein, B. Y.; Sharon, R.; Tarcic, N.; Naor, D.: Induction of antitumor reactive cells or suppressor cells by different molecular species isolated from the same non-immunogenic tumor. Immunobiology 163: 7 (1982).
805 Klein, E.; Vánky, F.; Masucci, M. G.; Bejarano, M. T.: Experimental and clinical data for the role of NK-cells in immunosurveillance. Behring Inst. Mitt. 74: 140 (1984).
806 Klostergaard, J.; Yamamoto, R. S.; Granger, G. A.: Human and murine lymphotoxins as a multicomponent system: Progress in purification of the human L component. Mol. Immunol. 17: 613 (1980).
807 Kluchareva, T. E.; Matveeva, V. A.; Deichman, G. I.: Sensitivity of TSTA and species-specific cell membrane antigens of tumor cells to glutaraldehyde treatment. Neoplasma 25: 273 (1978).
808 Knapp, W.: Monoclonal antibodies against differentiation antigens of myelopoiesis. Blut 45: 301 (1982).
809 Knop, J.: Effect of Vibrio cholerae neuraminidase on the mitogen response of T-lymphocytes. I. Enhancement of macrophage T-lymphocyte cooperation in concanavalin A-induced lymphocyte activation. Immunobiology 157: 474 (1980).
810 Knop, J.: Effect of Vibrio cholerae neuraminidase on the mitogen response of T-lymphocytes. II. Modulation of the lymphocyte response to macrophage released factors by neuraminidase. Immunobiology 157: 486 (1980).
811 Knop, J.; Sedlacek, H. H.; Seiler, F. R.: Stimulatory effect of Vibrio cholerae neuraminidase on the antibody response to various antigens. Immunology 34: 181 (1978).

812 Knop, J.; Stremmer, R.; Neumann, C.; De Maeyer, E.; Macher, E.: Interferon inhibits the suppressor T cell response of delayed-type hypersensitivity. Nature 296: 775 (1982).

813 Knuth, A.; Dippold, W.; Meyer zum Büschenfelde, K.-H.: Neue Aspekte der Tumorimmunologie: Die Bedeutung des T-Zellsystems. Dt. med. Wschr. 107: 1105 (1982).

814 Kobayashi, H.; Sendo, F.; Kaji, H.; Shirai, T.; Saito, H.; Takeichi, N.; Hosokawa, M.; Kodama, T.: Inhibition of transplanted rat tumors by immunization with identical tumor cells infected with Friend virus. J. natn. Cancer Inst. 44: 11 (1970).

815 Kobayashi, Y.; Sawada, J.; Osawa, T.: Isolation and characterization of an inhibitory glycopeptide against guinea pig lymphotoxin from the surface of L cells. Immunochemistry 15: 61 (1978).

816 Köhler, G.; Milstein, C.: Continuous cultures of fused cells secreting antibody of predefined specificity. Nature 256: 495 (1975).

817 Kölsch, E.; Mengersen, R.: Low numbers of tumor cells suppress the host immune system. Adv. Exp. Med. Biol. 66: 431 (1976).

818 Kojima, A.; Tamura, S. I.; Egashira, Y.: Regulatory role of suppressor cells stimulated with antigen in vitro and its possible interaction with macrophages. Immunology 37: 577 (1979).

819 Kolb, W. P.; Granger, G. A.: Lymphocyte in vitro cytotoxicity: characterization of human lymphotoxin. Proc. natn. Acad. Sci. 61: 1250 (1968).

820 Kolkovsky, P.; Bubenik, J.: Occurrence of tumours in mice after inoculation of rous sarcoma and antigenic changes in these tumours. Folia biol., Praha 10: 81 (1964).

821 Koo, G. C.; Hatzfeld, A.: Antigenic phenotype of mouse natural killer cells; in Herberman, Natural cell-mediated immunity against tumors, p. 105 (Academic Press, New York 1980).

822 Koprowski, H.; Love, R.; Koprowska, I.: Enhancement of susceptibility to viruses in neoplastic tissues. Tex. Rep. Biol. Med. 15: 559 (1957).

823 Koprowski, H.; Steplewski, Z.; Herlyn, D.; Herlyn, M.: Study of antibodies against human melanoma produced by somatic cell hybrids. Proc. natn. Acad. Sci. 75: 3405 (1978).

824 Koprowski, H.; Steplewski, Z.; Mitchell, K.; Herlyn, M.; Herlyn, D.; Fuhrer, P.: Colorectal carcinoma antigens detected by hybridoma antibodies. Somatic Cell Genet. 5: 957 (1979).

825 Koprowski, H.; Brockhaus, M.; Blaszczyk, M.; Magnani, J.; Steplewski, Z.; Ginsburg, V.: Lewis blood-type may affect the incidence of gastrointestinal cancer. Lancet i: 1332 (1982).

826 Koprowski, H.; Herlyn, D.; Lubeck, M.; DeFreitas, E.; Sears, H. F.: Human anti-idiotype antibodies in cancer patients: Is the modulation of the immune response beneficial for the patient? Proc. natn. Acad. Sci. 81: 216 (1984).

827 Korec, S.; Herberman, R. B.; Dean, J. H.; Cannon, G. B.: Cytostasis of tumor cell lines by human granulocytes. Cell. Immunol. 53: 104 (1980).

828 Koren, H. S.; Williams, M. S.: Natural killing and antibody-dependent cellular cytotoxicity are mediated by different mechanisms and by different cells. J. Immun. 121: 1956 (1978).

829 Koziner, B.; Denny, G. D.; McKenzie, S.; Clarkson, B. D.; Miller, D. A.; Evans,

R. L.: Analysis of T-cell differentiation antigens in acute lymphatic leukemia using monoclonal antibodies. Blood 60: 752 (1982).
830 Kreeftenberg, J. G.; de Jong, W.H.; Ettekoven, H.; Steerenberg, P. A.; Kruizinga, W.; van Norrle Jansen, L. M.; Sekhuis, J.; Ruitenberg, E. J.: Experimental screening to two BCG preparations produced according to different principles. Cancer Immunol. Immunother. 12: 21 (1981).
831 Kreider, J. W.; Bartlett, G. L.; Purnell, D. M.: Inconsistent response of B 16 melanoma to BCG immunotherapy. J. natn. Cancer Inst. 56: 803 (1976).
832 Kreider, J. W.; Bartlett, G. L.; Purnell, D. M.; Webb, S.: Destruction of regional lymph node metastases of rat mammary adenocarcinoma 13762A by treatment with C. parvum. Cancer Res. 38: 4522 (1978).
833 Kreider, J. W.; Batlett, G. L.; Boyer, C. M.; Purnell, D. M.: Condition for effective Bacillus Calmette-Guérin immunotherapy of postsurgical metastases of 13762A rat mammary adenocarcinoma. Cancer Res. 39: 987 (1979).
834 Krikorian, J. G.; Anderson, J. L.; Bieber, C. P.: Malignant neoplasms following cardiac transplantation. JAMA 240: 639 (1978).
835 Kristensen, E.; Langvad, E.; Reimann, R.: Humoral immunity in malignant skin melanoma. Isolation of melanoma specific IgG from melanoma metastases. Eur. J. Cancer 12: 945 (1976).
836 Krolick, K. A.; Yuan, D.; Vitetta, E. S.: Specific killing of a human breast carcinoma cell line by a monoclonal antibody coupled to the A-chain of ricin. Cancer Immunol. Immunother. 12: 39 (1981).
837 Krolick, K. A.; Uhr, J. W.; Slavin, S.; Vitetta, E. S.: In vivo therapy of a murine B cell tumor (BCL_1) using antibody ricin A chain immunotoxins. J. exp. Med. 155: 1797 (1982).
838 Krupey, J.; Gold, P.; Freedman, S. O.: Physicochemical studies of the carcinoembryonic antigens of the human digestive system. J. exp. Med. 128: 387 (1968).
839 Kudo, T.; Aoki, T.; Morrison, J. L.: Stabilization of antigens on surfaces of malignant cells by formalin treatment. J. natn. Cancer Inst. 52: 1553 (1974).
840 Kull, F. C., Jr.; Jacobs, S.; Cuatrecasas, P.: Cellular receptor for ^{125}I-labeled tumor necrosis factor: Specific binding, affinity labeling, and relationship to sensitivity. Proc. natn. Acad. Sci. USA 82: 5756 (1985).
841 Kumar, R. K.; Lykke, A. W. J.; Penny, R.: Immunosuppression associated with SJL/J murine lymphoma. II. Characterisation of a plasma suppressor factor in tumor-bearing mice. J. natn. Cancer Inst. 67: 1277 (1981).
842 Kumar, R. K.; Penny, R.: Escape of tumours from immunological destruction. Pathology 14: 173 (1982).
843 Kurland, J. I.; Bockman, R.: Prostaglandin E production by human blood monocytes and mouse peritoneal macrophages. J. exp. Med. 147: 952 (1978).
844 Kurland, J. I.; Bockman, R. S.; Broxmeyer, H. E.; Moore, M. A. S.: Limitation of excessive myelopoiesis by the instrinsic modulation of macrophage-derived prostaglandin E. Science 199: 552 (1978).
845 Kurland, J. I.; Pelus, L. M.; Ralph, P.; Bockman, R. S.; Moore, M. A. S.: Induction of prostaglandin E synthesis in normal and neoplastic macrophages: Role of colony-stimulating factor(s) distinct from effects on myeloid progenitor cell proliferation. Proc. natn. Acad. Sci. USA 76: 2326 (1979).

846 Kurth, R.: Grenzen und Möglichkeiten der Immuntherapie von Tumoren. Naturwissenschaften *65:* 180 (1978).
847 Kurth, R.; Fenyö, E. M.; Klein, E.; Essex, M.: Cell-surface antigens induced by RNA tumour viruses. Nature *279:* 197 (1979).
848 Lachman, L. B.; Hacker, M. P.; Handschumacher, R. E.: Partial purification of human lymphocyte-activating factor (LAF) by ultrafiltration and electrophoretic techniques. J. Immun. *119:* 2019 (1977).
849 Lacour, F.; Delage, G.; Chianale, C.: Reduced incidence of spontaneous mammary tumors in C3H/He mice after treatment with polyadenylate-polyuridylate. Science *187:* 256 (1975).
850 Lai, A.; Fat, R. M.; van Furth, R.: In vitro synthesis of some complement components (Clq, C3 and C4) by lymphoid tissues and circulating leukocytes in man. Immunology *28:* 359 (1975).
851 Lamerz, R.; Fateh-Moghadam, A.: Carcinofetale Antigene. I. Alpha-Fetoprotein. Klin. Wschr. *53:* 147 (1975).
852 Lamerz, R.; Fateh-Moghadam, A.: Carcinofetale Antigene. II. Carcinoembryonales Antigen (CEA). Klin. Wschr. *53:* 193 (1975).
853 Lamerz, R.; Fateh-Moghadam, A.: Carcinofetale Antigene. III. Andere carcinofetale Antigene. Klin. Wschr. *53:* 403 (1975).
854 Lamm, D. L.; Thor, D. E.; Harris, S. C.; Reyna, J. A.; Stogdill, V. D.; Radwin, H. M.: Bacillus Calmette-Guérin immunotherapy of superficial bladder cancer. J. Urol. *124:* 38 (1980).
855 Larizza, L.; Schirrmacher, V.; Graf, L.; Pflüger, E.; Peres-Martinez, M.; Stöhr, M.: Suggestive evidence that the highly metastatic variant ESb of the T-cell lymphoma Eb is derived from spontaneous fusion with a host macrophage. Int. J. Cancer *34:* 699 (1984).
856 Larsson, A.; Pisarri-Salsano, S.; Öhlander, C.; Natvig, J. V.; Perlmann, P.: Destruction of dextran coated target cells by normal human lymphocytes and monocytes: Induction by a human anti-dextran serum with IgG antibodies restricted to the IgG2 subclass. Scand. J. Immunol. *4:* 241 (1975).
857 Laszlo, J.; Buckley, C. E.; Amos, D. B.: Infusion of isologous immune plasma in chronic lymphocytic leukaemia. Blood *31:* 104 (1968).
858 Law, I.: Some biologic, immunologic, and morphologic effects in mice after infection with a murine sarcoma virus. I. Biologic and immunologic studies. J. natn. Cancer. Inst. *40:* 1101 (1968).
859 Lawrence, H. S.: The transfer of generalized cutaneous hypersensitivity of the delayed tuberculin type in man by means of constituents of disrupted leukocytes. J. clin. Invest. *33:* 951 (1954).
860 Lazarides, E.: Intermediate filaments: A chemically heterogeneous, developmentally regulated class of proteins. Annu. Rev. Biochem. *51:* 219 (1982).
861 Le, J.; Yip, Y. K.; Vilček, J.: Cytolytic activity of interferon-gamma and its synergism with 5-fluorouracil. Int. J. Cancer *34:* 495 (1984).
862 Lechler, R. I.; Batchelor, J. R.: Restoration of immunogenicity to passenger cell depleted kidney allografts by the addition of donor strain dendritic cells. J. exp. Med. *155:* 31 (1982).
863 Lee, C. S.; Chen, S. H.; Lin, T. Y.: Inhibition of leukocyte migration by tumor-associated antigen in soluble extracts of human hepatoma. Cancer Res. *37:* 918 (1977).

864 Lemonnier, F.; Burakoff, S. J.; Germain, R. N.; Benacerraf, B.: Cytolytic thymus derived lymphocytes specific for allogeneic stimulator cells crossreact with chemically modified syngeneic cells. Proc. natn. Acad. Sci. USA 74: 1229 (1977).
865 Lerner, R. A.: Tapping the immunological repertoire to produce antibodies of predetermined specificity. Nature 299: 592 (1982).
866 Lespinats, G.: Tumor specific humoral antibodies against plasma cell tumors in immunized Balb/c mice. J. natn. Cancer Inst. 45: 845 (1970).
867 Leung-Tack, J.; Maillard, J.; Voisin, G. A.: Chemotaxis for polymorphonuclear leukocytes induced by soluble antigen-antibody complexes. Immunol. 33: 937 (1977).
868 Levine, B. B.; Ojeda, A.; Benacerraf, B.: Studies on artificial antigens. III. The genetic control of the immune response to hapten-poly-L-lysine conjugates in guinea pigs. J. exp. Med. 118: 953 (1963).
869 Levine, B. B.; Benacerraf, B.: Genetic control in guinea pigs of immune response to conjugates of haptens and poly-L-lysine. Science 147: 517 (1965).
870 Levy, M. H.; Wheelock, E. F.: The role of macrophages in defense against neoplastic disease. Adv. Cancer Res. 20: 131 (1974).
871 Levy, M. H.; Wheelock, E. F.: Effects of intravenous silica on immune and non-immune functions of the murine host. J. Immun. 115: 41 (1975).
872 Levy, L.; Chia, D.; Barnett, E. V.: Effect of pulse dose cyclophosphamide on the anamnestic immune response in NZB/W mice. Agents and Actions 8: 644 (1978).
873 Levy, R.; Miller, R. A.: Tumor therapy with monoclonal antibodies. Fed. Proc. 42: 2650 (1983).
874 Liabeuf, A.; LeBorgne de Kaouel, C.; Kourilsky, F. M.; Malissen, B., Manuel, Y.; Sanderson, A. R.: An antigenic determinant of human β_2-microglobulin masked by the association with HLA heavy chains at the cell surface: Analysis using monoclonal antibodies. J. Immunol. 127: 1542 (1981).
875 Liao, S. K.; Clarke, B. J.; Kwong, P. C.: Common neuroectodermal antigens on human melanoma, neuroblastoma, retinoblastoma, glioblastoma, and fetal brain revealed by the hybridoma antibodies raised against melanoma cells. Eur. J. Immunol. 11: 450 (1981).
876 Likhite, V. V.: Suppression of the incidence of death with spontaneous tumours in DBA/2 mice after Corynebacterium parvum-mediated rejection of syngeneic tumours. Nature 259: 397 (1976).
877 Lillihoj, H.-S.; Choe, B.-K.; Rose, N. R.: Monoclonal anti-human prostatic acid phosphatase antibodies. Mol. Immunol. 19: 1199 (1982).
878 Lin, J. S. L.; Huber, N.; Murphy, W. H.: Immunization of C58 mice to line I_b leukemia. Cancer Res. 29: 2157 (1969).
879 Lin, J. S.; Murphy, W. H.: Dependence of immunity to IB leukemia on an adjuvant effect of immunizing cell preparations. Cancer Res. 29: 2163 (1969).
880 Lindahl, P.; Leary, P.; Gresser, I.: Enhancement by interferon of the specific cytotoxicity of sensitized lymphocytes. Proc. natn. Acad. Sci. USA 69: 721 (1972).
881 Lindahl, P.; Leary, P.; Gresser, I.: Enhancement by interferon of the expression of surface antigens on murine leukemia L1210-cells. Proc. natn. Acad. Sci. USA 70: 2785 (1973).
882 Lindahl, P.; Leary, P.; Gresser, I.: Enhancement of the expression of histocompatibility antigens of mouse lymphoid cells by interferon in vitro. Eur. J. Immunol. 4: 779 (1974).

883 Lindahl-Kiessling, K.; Peterson, R. D. A.: The mechanism of phytohemagglutinin action. III. Stimulation of lymphocytes by allogeneic lymphocytes and phytohemagglutinin. Expl. Cell Res. *55:* 85 (1969).
884 Lindahl-Magnussen, P.; Leary, P.; Gresser, I.: Interferon and cell division. VI. Inhibitory effect of interferon on the multiplication of mouse embryo and mouse kidney cells in primary cultures. Proc. Soc. exp. Biol. Med. *138:* 1044 (1971).
885 Lindahl-Magnussen, P.; Leary, P.; Gresser, I.: Interferon inhibits DNA synthesis induced in mouse lymphocyte suspensions by phytohaemagglutinin or by allogeneic cells. Nature new Biol. *237:* 120 (1972).
886 Lindemalm, C. S.; Killander, A.; Björkholm, M.; Brenning, G.; Engstedt, L.; Franzen, L.; Gahrton, G.; Gullbring, B.; Holm, G.; Höglund, S.; Hörnsten, P.; Jameson, S.; Killander, D.; Klein, E.; Lantz, B.; Lockner, D.; Lönnqvist, B.; Mellstedt, H.; Palmband, J.; Pauli, C.; Reizenstein, P.; Simonsson, B.; Skarberg, K. P.; Uden, A. M.; Vanky, F.; Wadman, B.: Adjuvant immunotherapy in acute nonlymphocytic leukemia. Cancer Immunol. Immunother. *4:* 179 (1978).
887 Lindenmann, J.: Viral oncolysis with host survival. Proc. Soc. exp. Biol. Med. *113:* 85 (1963).
888 Lindenmann, J.; Klein, P. A.: Immunological aspects of viral oncolysis. Recent Results Cancer Res. *9:* 1 (1967).
889 Lindenmann, J.: Immunogenicity of oncolysates obtained from Ehrlich ascites tumours infected with vesicular stomatitis virus. Arch. ges. Virusforsch. *31:* 61 (1970).
890 Lindenmann, J.: Viruses as immunological adjuvants in cancer. Biochim. biophys. Acta *49:* 355 (1974).
891 Linker-Israeli, M.; Billing, R. J.; Foon, K. A.; Terasaki, P. I.: Monoclonal antibodies reactive with acute myelogenous leukemia cells. J. Immunol. *127:* 2473 (1981).
892 Littman, M. L.; Kim, Y. C.; Suk, D.: Immunization of mice to sarcoma 180 and Ehrlich carcinoma with ultraviolet-killed tumor vaccine. Proc. Soc. exp. Biol. Med. *127:* 7 (1968).
893 Lobuglio, A. F.: Effect of neoplasia on human macrophage activity. J. Lab. clin. Med. *76:* 888 (1970).
894 Lohmann-Matthes, M. L.; Domzig, W.; Zähringer, M.; Lang, H.: K cell and NK cell like activity of macrophage precursor cells. Behring Inst. Mitt. *65:* 26 (1980).
895 Loitta, L. A.; Kleinerman, J.; Saidel, G. M.: Mechanism of bacillus Calmette-Guérin induced suppression of metastases in a poorly immunogenic fibrosarcoma. Cancer Res. *36:* 3255 (1976).
896 Londner, M. V.; Morini, J. C.; Font, M. T.; Rabasa, S. L.: RNA-induced immunity against a rat sarcoma. Experientia *24:* 598 (1968).
897 Longo, D. L.; Matis, L. A.; Schwartz, R. H.: Insights into immune response gene function from experiments with chimeric animals. CRC crit. Rev. Immunol. *2:* 83 (1981).
898 Loop, S. M.; Nishiyama, K.; Hellström, I.; Woodbury, R. G.; Brown, J. P.; Hellström, K. E.: Two human tumor-associated antigens, p155 and p210, detected by monoclonal antibodies. Int. J. Cancer *27:* 775 (1981).
899 Lopez-Berestein, G.; Mehta, K.; Mehta, R.; Juliano, R. L.; Hersh, E. M.: The activation of human monocytes by liposome-encapsulated muramyl dipeptide analogues. J. Immunol. *130:* 1500 (1983).

900 Loring, M.; Schlesinger, M.: The 0 antigenicity of lymphoid organs of mice bearing Ehrlich ascites tumor. Cancer Res. *30:* 2204 (1970).
901 Lotzovà, E.: C. parvum-mediated suppression of the phenomenon of natural killing and its analysis; in Herberman, Natural cell-mediated immunity against tumors, p. 735 (Academic Press, New York 1980).
902 Loutit, J. F.; Townsend, K. M. S.; Knowles, J. F.: Tumour surveillance in beige mice. Nature *286:* 66 (1980).
903 Lovchik, J.; Hong, R.: Characterization of effector and target cell populations in antibody-dependent cell-mediated cytolysis. Fed. Proc. *33:* 780 (1974).
904 Lovchik, J. C.; Hong, R.: Antibody-dependent cell-mediated cytolysis (ADCC): Analyses and projections; in Kallós, Waksman, de Weck, Progress in allergy, vol. 22, p. 1 (Karger, Basel 1977).
905 Lubet, R. A.; Carlson, D. E.: Therapy of the murine plasmacytoma MOPC 104E: role of the immune response. J. natn. Cancer Inst. *61:* 897 (1978).
906 Lüben, G.; Sedlacek, H. H.; Seiler, F. R.: Quantitative experiments on the cell membrane binding of neuraminidase. Behring Inst. Mitt. *59:* 30 (1976).
907 Lumsden, T.: Tumor immunity: The effects of the eu- and pseudo-globulin fractions of anti-cancer sera on tissue culture. J. Path. Bact. *34:* 349 (1931).
908 Lundgren, G.; Zukoski, C. F.; Möller, G.: Differential effects of human granulocytes and lymphocytes on human fibroblasts in vitro. Clin. exp. Immunol. *3:* 817 (1968).
909 Lurie, M. G.: Studies on the mechanisms of immunity in tuberculosis. The fate of tubercle bacilli ingested by mononuclear phagocytes derived from normal and immunized animals. J. exp. Med. *75:* 247 (1942).
910 McDermott, R. P.: Chess, L.; Schlossman, S. F.: Immunological functions of isolated human lymphocyte subpopulations. V. Isolation and functional analysis of a surface Ig-negative, E-rosette-negative subset. Clin. Immunol. Immunother. *4:* 415 (1975).
911 Mackaness, G. B.: Cellular resistance to infection. J. exp. Med. *116:* 381 (1962).
912 Mackaness, G. B.: The immunological basis of acquired cellular resistance. J. exp. Med. *120:* 105 (1964).
913 Mackaness, G. B.; Lagrange, P. H.; Miller, T. E.; Ishibashi, T.: The formation of activated T-cells; in Wagner et al., Activation of macrophages, p. 193 (Excerpta Medica, Amsterdam 1974).
914 MacLennan, I. C. M.: Antibody in the induction and inhibition of lymphocyte cytotoxicity. Transplant. Rev. *13:* 67 (1972).
915 MacLennan, I. C. M.: Competition for receptors for immunoglobulin on cytotoxic lymphocytes. Clin. exp. Immunol. *10:* 275 (1972).
916 MacLennan, I. C. M.; Howard, A.; Gotch, F. M.; Quie, P. G.: Effector activating determinants on IgG. I. The distribution and factors influencing the display of complement, neutrophil and cytotoxic B-cell determinants on human IgG subclasses. Immunology *25:* 459 (1973).
917 Margarey, C. J.; Baum, M.: Reticuloendothelial activity in humans with cancer. Br. J. Surg. *57:* 748 (1970).
918 Magnani, J. L.; Brockhaus, M.; Smith, D. F.; Ginsburg, V.; Blaszczyk, M.; Mitchell, K. F.; Steplewski, Z.; Koprowski, H.: A monosialoganglioside is a monoclonal antibody-defined antigen of colon carcinoma. Science *212:* 55 (1981).

919 Mahoney, M. J.; Leighton, J.: The inflammatory response to a foreign body within transplantable tumors. Cancer Res. 22: 334 (1962).
920 Makidono, R. et al.: Enhanced development of metastatic foci in thymectomised, irradiated and bone marrow-reconstituted mice. Gann 67: 645 (1976).
921 Månsson, J.-E.; Fredman, P.; Nilsson, O.; Lindholm, L.; Holmgren, J.; Svennerholm, L.: Chemical structure of carcinoma ganglioside antigens defined by monoclonal antibody C-50 and some allied gangliosides of human pancreatic adenocarcinoma. Biochim. biophys. Acta 834: 110 (1985).
922 Martin, W. J.; Wunderlich, J. R.; Fletcher, F.; Inman, J. K.: Enhanced immunogenicity of chemically-coated syngeneic tumor cells. Proc. natn. Acad. Sci. USA 68: 469 (1971).
923 Martin, W. J.; Gipson, T. G.; Conliffe, M.; Friedman, R. J.; Dove, L., Rice, J. M.: Common tumor-associated transplantation alloantigen detected on a proportion of lung tumors induced transplacentally in several strains of mice. Transplantation 24: 294 (1977).
924 Martin, W. J.; Gipson, T. G.; Conliffe, M. A.; Cotton, W. G.; Dove, L. F.; Rice, J. M.: Histocompatibility difference between C3HfeB/HeN and C3H/HeN mice: tumour-induced in C3HFeB/HeN mice expresses C3H/HeN-associated alloantigen. J. Immunogenet. 5: 225 (1978).
925 Martin, W. J.; Imamura, M.: Variable expression of histocompatibility antigens on tumor cells. Cancer Immunol. Immunother. 8: 219 (1980).
926 Mason, D. W.; Pugh, C. W.; Webb, M.: The rat mixed lymphocyte reaction: roles of a dendritic cell in intestinal lymph and T-cell subsets defined by monoclonal antibodies. Immunology 44: 75 (1981).
927 Masui, H.; Kawamoto, T.; Sato, J. D.; Wolf, B.; Sato, G.; Mendelsohn, J.: Growth inhibition of human tumor cells in athymic mice by anti-epidermal growth factor receptor monoclonal antibodies. Cancer Res. 44: 1002 (1984).
928 Mathé, G.: Successful allogeneic bone-marrow transplantation in man: Chimaerism, induced by specific tolerance and possible anti-leukaemic effects. Blood 25: 179 (1965).
929 Mathé, G.: Immunothérapie active de la leucémie L1210 appliquée après la geffe tumorale. Revue fr. Étud. clin. biol. 13: 881 (1968).
930 Mathé, G.: Immunotherapy in the treatment of acute lymphoid leukemia. Hosp. Pract. 6: 43 (1971).
931 Mathé, G.; Loc, T. B.; Bernard, J.: Effet sur la leucémie L1210 de la souris d'une combinaison par diazotation d'A-méthopterine et de -globulines de hamsters porteurs de cette leucémie par heterogreffe. Compt. Rend. Acad. Sci. 246: 1626 (1958).
932 Mathé, G.; Amiel, J. L.; Schwarzenberg, L.; Schneider, M.; Cattan, A.; Schlumberger, J. R.; Hayat, M.; De Vassal, F.: Active immunotherapy for acute lymphoblastic leukaemia. Lancet i: 697 (1969).
933 Mathé, G.; Pouillart, P.; Lapeyrague, F.: Active immunotherapy of L1210 leukaemia applied after the graft of tumour cells. Br. J. Cancer 23: 814 (1969).
934 Mathé, G.; Halle-Pannenko, O.; Bourut, C.: Active immunotherapy of AKR mice with spontaneous leukemia. Revue eur. Étud. Clin. Biol. 17: 997 (1973).
935 Mathé, G.; Halle-Pannenko, O.; Bourut, C.: Effectiveness of murine leukemia chemotherapy according to the immune state. Cancer Immunol. Immunother. 2: 139 (1977).

936 Mathé, G.; Florentin, I.; Ohnoi, L.; Bruley-Rosset, M.; Schulz, J.; Kiger, N.; Orbach-Arboiys, S.; Schwarzenberg, L.; Pouillart, P.; De Vassal, F.: Pharmacologic factors and manipulation of immunity systemic adjuvants in cancer therapy. Cancer Treat. Rep. *62:* 1613 (1978).

937 Mattes, M. J.; Sharrow, S. O.; Herberman, R. B.; Holden, H. T.: Identification and separation of Thy-1 positive mouse spleen cells active in natural cytotoxicity and antibody-dependent cell-mediated cytotoxicity. J. Immunol. *123:* 2851 (1979).

938 Mayhew, E.; Weiss, L.: Ribonucleic acid at the periphery of different cell types and effect of growth rate on ionogenic groups in the periphery cultured cells. Expl Cell Res. *50:* 441 (1968).

939 Mazauric, T.; Mitchell, K. F.; Letchworth, G. J., III.; Koprowski, H.; Steplewski, Z.: Monoclonal antibody-defined human lung cell surface protein antigens. Cancer Res. *42:* 150 (1982).

940 McAllister, R. M.; Grunmeier, P. W.; Coriell, L. L.: The effects of heterologous immune serum upon HeLa cells in vitro and rat-HeLa tumors in vivo. J. natn. Cancer Inst. *21:* 541 (1958).

941 McCabe, R. P.; Evans, C. H.: Induction of lymphotoxin secretion from human, hamster and guinea pig lymphocytes by physiologic concentrations of human thrombin. Fed. Proc. *42:* 682 (1983).

942 McCarthy, W. H.; Cotton, G.; Carlon, A.; Milton, G. W.; Kossard, S.: Immunotherapy of malignant melanoma: A clinical trial. Cancer *32:* 97 (1973).

943 McCollester, D. L.: Isolation of Meth A cell surface membranes possessing tumor-specific transplantation antigen activity. Cancer Res. *30:* 2832 (1970).

944 McCoy, J. L.; Jerome, L. F.; Anderson, C.; Cannon, G. B.; Alford, T. C.; Connor, R. J.; Oldham, R. K.; Herberman, R. B.: Leukocyte migration inhibition by soluble extracts of MCF-7 tissue culture cell line derived from breast carcinoma. J. natn. Cancer Inst. *57:* 1045 (1976).

945 McCredie, J. A.; MacDonald, H. R.: Antibody-dependent cellular cytotoxicity in cancer patients: Lack of prognostic value. Br. J. Cancer *41:* 880 (1980).

946 McDevitt, H. O.; Sela, M.: Genetic control of the antibody response. I. Demonstration of determinant-specific differences in response to synthetic polypeptide antigens in two strains of inbred mice. J. exp. Med. *122:* 517 (1965).

947 McDevitt, H. O.; Chinitz, A.: Genetic control of the antibody response: relationship between immune response and histocompatibility (H-2) type. Science *163:* 1207 (1969).

948 McGuire, R. L.; Fox, R. A.: Suppression of delayed hypersensitivity by the depletion of circulating monocytes. Immunology *38:* 157 (1979).

949 McHowell, J.; Ishmael, J.; Tandy, J.; Hughes, I. B.: A 6 year survey of tumours of dogs and cats removed surgically in private practice. J. small Anim. Pract. *11:* 793 (1970).

950 McIllmurray, M. B.; Embleton, M. J.; Reeves, W. G.: Controlled trial of active immunotherapy in management of stage II B malignant melanoma. Br. Med. J. *i:* 540 (1977).

951 McKenzie, I. F. C.; Clarke, A. E.; Parish, C. R.: Ia antigenic specificities are oligosaccharide in nature: Hapten inhibition studies. J. exp. Med. *145:* 1039 (1977).

952 McKhann, C. F.: The effect of X-ray on the antigenicity of donor cells in transplantation immunity. J. Immun. *92:* 811 (1964).

953 McMichael, H.: Inhibition by methylprednisolone of regression of the shope rabbit papilloma. J. natn. Cancer Inst. *39:* 55 (1967).
954 McNeill, T. A.; Gresser, I.: Inhibition of hemopoietic colony growth by interferon preparations from different sources. Nature new Biol. *244:* 173 (1973).
955 McQuiddy, P.; Lilien, J. E.: The binding of exogenously added neuraminidase to cells and tissues in culture. Biochim. biophys. Acta *291:* 774 (1973).
956 Medina, D.; Heppner, G.: Cell-mediated immunostimulation induced by mammary tumor virusfree Balb/c mammary tumors. Nature, Lond. *242:* 329 (1973).
957 Meltzer, M. S.; Leonard, E. J.; Rapp, H. J.; Borsos, T.: Tumor-specific antigen solubilized by hypertonic potassium chloride. J. natn. Cancer Inst. *47:* 703 (1971).
958 Meltzer, M. S.; Oppenheim, J. J.; Lettman, B. H.; Leonard, E. J.; Rapp, H. F.: Cell-mediated tumor immunity measured in vitro and in vivo with soluble tumor-specific antigens. J. natn. Cancer Inst. *49:* 727 (1972).
959 Meltzer, M. S.; Leonard, E. J.: Enhanced tumour growth in animals pre-treated with complete Freund's adjuvant. J. natn. Cancer Inst. *50:* 209 (1973).
960 Meltzer, M. S.; Tucker, R. W.; Sanford, K. K.: Interaction of BCG-activated macrophages with neoplastic and non-neoplastic cell lines in vitro: Quantitation of the cytotoxic reaction by release of tritiated thymidine from prelabeled target cells. J. natn. Cancer Inst. *54:* 1177 (1975).
961 Meltzer, M. S.; Oppenheim, J. J.: Bidirectional amplification of macrophage-lymphocyte interactions: Enhanced LAF production by activated adherent mouse peritoneal cells. J. Immun. *118:* 77 (1977).
962 Meltzer, M. S.; Stevenson, M. M.: Macrophage function in tumour-bearing mice: Tumoricidal and chemotactic response of macrophages activated by infection with mycobacterium bovis strain BCG. J. Immun. *118:* 2176 (1977).
963 Menard, S.; Colnaghi, M. I.; Della Porta, G.: In vitro demonstration of tumor-specific common antigens and embryonal antigens in murine fibrosarcomas induced by 7-12-dimethyl-benz-A-anthracene. Cancer Res. *33:* 478 (1973).
964 Mendelsohn, J. et al.: Effects of anti-EGF receptor monoclonal antibodies upon human tumor cells. J. cell. Biochem., suppl. 9A, p. 65 (1985).
965 Meo, T.; David, C. S.; Rijnbeek, A. M.; Nabholz, M.; Miggiano, V. C.; Shreffler, D. C.: Inhibition of mouse MLR by anti-Ia sera. Transplant. Proc. *7:* suppl. 1, p. 127 (1975).
966 Mesa-Tejada, R.; Oster, M. W.; Fenoglio, C. M.; Magidson, J.; Spiegelman, S.: Diagnosis of primary breast carcinoma through immunohistochemical detection of antigen related to mouse mammary tumor virus in metastatic lesions: A report of two cases. Cancer *49:* 261 (1982).
967 Meschini, A.; Invernizzi, G.; Parmiani, G.: Expression of alien specificities on a chemically induced Balb/c fibrosarcoma. Int. J. Cancer *20:* 271 (1977).
968 Metzgar, R. S.; Olenick, S. R.: The study of normal and malignant cell antigens by mixed agglutination. Cancer Res. *28:* 1366 (1968).
969 Metzgar, R. S.; Mohanakumar, T.; Miller, D. S.: Antigens specific for human lymphocytic and myeloid leukemia cells: Detection of nonhuman primate antiserums. Science *178:* 986 (1972).
970 Metzgar, R. S.; Gaillard, M. T.; Levine, S. J.; Tuck, F. L.; Bossen, E. H.; Borewitz, M. J.: Antigens of human pancreatic adenocarcinoma cells defined by murine monoclonal antibodies. Cancer Res. *42:* 601 (1982).

971 Metzger, Z.; Hoffeld, J. T.; Oppenheim, J. J.: Macrophage-mediated suppression. I. Evidence for participation of both hydrogen peroxide and prostaglandins in suppression of murine lymphocyte proliferation. J. Immun. *124:* 983 (1980).
972 Michaelides, M. C.; Sato, N.; Wallack, M. K.: Screening for monoclonal antibodies to cell surface antigens using the ELISA test on terasaki plates. J. immunol. Methods *58:* 267 (1983).
973 Middle, J. G.; Embleton, M. J.: Naturally arising tumors of the inbred WAB/not rat strain. II. Immunogenicity of transplanted tumors. J. natn. Cancer Inst. *67:* 637 (1981).
974 Mikulski, S. M.; McGuire, W. P.; Loure, A. C.: Immunotherapy of lung cancer. I. Review of clinical trials in non small cell histological types. Cancer Treat. Rev. *6:* 177 (1979).
975 Milas, L.; Hunter, N.; Basic, I.; Withers, H. R.: Protection by C. granulosum against radiation-induced enhancement of artificial metastasis of a murine fibrosarcoma. J. natn. Cancer Inst. *52:* 1875 (1974).
976 Millar, D.; Salaman, J. R.: The sensitising properties of renal perfusate. Transplantation *18:* 251 (1974).
977 Milleck, J.; Pasternak, G.: Zur in vivo Wirksamkeit heterologer Antiseren vom Kaninchen gegen Leukämien der Maus. Arch. Geschwulstforsch. *42:* 192 (1973).
978 Miller, J. F. A. P.: Immunological significance of the thymus of the adult mouse. Nature, Lond. *195:* 1318 (1962).
979 Miller, J. F. A. P.; Vadas, M. A.; Whitelaw, A.; Gamble, J.: H-2 gene complex restricts transfer of delayed-type hypersensitivity in mice. Proc. natn. Acad. Sci. USA *72:* 5095 (1975).
980 Miller, C. W.; DeBlasi, R. F.; Fisher, S. J.: Immunological studies in murine osteosarcoma. J. Bone Jt Surg. *58:* 312 (1976).
981 Miller, J. F. A. P.; Vadas, M. A.; Whitelaw, A.; Gamble, J.: Role of major histocompatibility complex gene products in delayed-type hypersensitivity. Proc. natn. Acad. Sci. USA *73:* 2486 (1976).
982 Miller, J. F. A. P.; Gamble, J.; Mottram, P.; Smith, F. I.: Influence of thymus genotype on acquisition of responsiveness in delayed-type hypersensitivity. Scand. J. Immunol. *9:* 29 (1979).
983 Miller, R. A.; Maloney, D. G.; Warnke, R.; Levy, R.: Treatment of B-cell lymphoma with monoclonal anti-idiotype antibody. New Engl. J. Med. *306:* 517 (1982).
984 Mills, G.; Monticone, V.; Paetkau, V.: The role of macrophages in thymocyte mitogenesis. J. Immun. *117:* 1325 (1976).
985 Minna, J. D.; Cuttitta, FD.; Rosen, S.; Bunn, P. A., Jr.; Carney, D. N.; Gazdar, A. F.; Krasnow, S.: Methods for production of monoclonal antibodies with specificity for human lung cancer cells. In Vitro *17:* 1058 (1981).
986 Mitchell, K. F.; Fuhrer, J. P.; Steplewski, Z.; Koprowski, H.: Biochemical characterization of human melanoma cell surfaces: Dissection with monoclonal antibodies. Proc. natn. Acad. Sci. *77:* 7287 (1980).
987 Mitchison, N. A.: Immunologic approach to cancer. Transplant. Proc. *2:* 92 (1970).
988 Mitchison, N. A.: The carrier effect in the secondary response to hapten-protein conjugates. II. Cellular cooperation. Eur. J. Immunol. *1:* 18 (1971).
989 Miyoshi, I.; Kubonishi, I.; Yoshimoto, S.; Akagi, T.; Ohisuki, Y.; Shiraishi, Y.; Nagata, K.; Hinuma, Y.: Detection of type-C virus particles in a cord T-cell line

derived by cocultivation of normal human cord leukocytes and human leukemic T-cells. Nature, Lond. *296:* 770 (1981).

990 Mizel, S. B.; Oppenheim, J. J.; Rosenstreich, D. L.: Characterization of LAF produced by the macrophage cell line, P388D$_1$. I. Enhancement of LAF production by activated T lymphocytes. J. Immun. *120:* 1497 (1978).

991 Moayeri, H.; Takita, H.; Sokal, J. E.: Immunotherapy of lung cancer with Corynebacterium parvum. Cancer Immunol. Immunother. *6:* 223 (1979).

992 Möller, G.: Demonstration of mouse isoantigens at the cellular level by the fluorescent antibody technique. J. exp. Med. *114:* 415 (1961).

993 Möller, E.: Antagonistic effects of humoral isoantibodies on the in vitro cytotoxicity of immune lymphoid cells. J. exp. Med. *122:* 11 (1965).

994 Möller, G.: β_2-microglobulin and HLA-antigen. Transplantation *21:* 1 (1974).

995 Mohr, E.; Schramm, G.: Reinigung und Charakterisierung der Neuraminidase aus Vibrio cholerae. Z. Naturf. *15 B:* 568 (1960).

996 Moll, R.; Franke, W. W.; Schiller, D. L.; Geiger, B.; Krepler, R.: The catalog of human cytokeratins: Patterns of expression in normal epithelia, tumors and cultured cells. Cell *31:* 11 (1982).

997 Moolten, F. L.; Capparell, N. J.; Zajdel, S. H.; Cooperband, S. R.: Anti-tumor effects of antibody-diphteria toxin conjugates. II. Immunotherapy with conjugates directed against tumor antigens induced by Simian virus 40. J. natn. Cancer Inst. *55:* 473 (1975).

998 Moore, R. N.; Oppenheim, J. J.; Farrar, J. J.; Carter, C. S., Jr.; Waheed, A.; Shadduck, R. K.: Production of LAF (IL1) by macrophages activated with colony stimulating factors. J. Immun. *125:* 1302 (1980).

999 Moretta, L.; Webb, S. R.; Grossi, C. E.; Lydyard, P. M.; Cooper, M. D.: Functional analysis of two human T cell subpopulations: Help and suppression of B cell responses by T cells bearing receptors for IgM and IgG. J. exp. Med. *146:* 184 (1977).

1000 Morgan, E. L.; Tempelis, C. H.: The role of antigen-antibody complexes in mediating immunologic unresponsiveness in the chicken. J. Immun. *119:* 1293 (1977).

1001 Morgan, E. L.; Tempelis, C. H.: The requirement for the Fc portion of antibody in antigen-antibody complex-mediated suppression. J. Immun. *120:* 1669 (1978).

1002 Morton, D. L.; Malmgren, R. A.; Holmes, E. C.; Ketcham, A.: Demonstration of antibodies against human malignant melanoma by immunofluorescence. Surgery *64:* 233 (1968).

1003 Morton, D. L.; Miller, G. F.; Wood, D. A.: Demonstration of tumorspecific immunity against antigens unrelated to the mammary tumor virus in spontaneous mammary adenocarcinomas. J. natn. Cancer Inst. *42:* 289 (1969).

1004 Morton, D. L.; Eilber, F. R.; Malmgren, R. A.; Wood, W. C.: Immunological factors which influence response to immunotherapy in malignant melanoma. Surgery, St. Louis *68:* 158 (1970).

1005 Morton, D. L.: Immunological studies with human neoplasms. J. reticuloendoth. Soc. *10:* 137 (1971).

1006 Motta, R.: Passive immunotherapy of leukemia and other cancer. Adv. Cancer Res. *14:* 161 (1971).

1007 Moyers, C.; Dröge, W.: Antigen presenting cells for cytotoxic T lymphocyte precursor cells. Behring Inst. Mitt. *70:* 106 (1982).

1008 MRC Report: Immunotherapy of acute myeloid leukemia. Br. J. Cancer *37:* 1 (1978).
1009 Müller, M.; Irmscher, J.; Fischer, R.; Grossmann, H.: Immunologisches Tumorprofil. Dt. GesundhWes. *39:* 1836 (1975).
1010 Müller-Eberhard, H. J.: Complement. Annu. Rev. Biochem. *44:* 697 (1975).
1011 Muhrer, K.-H.; Stambolis, C.: Nachweis immunologischer Kriterien im Elektrophorese-Mobilitäts-Test? Tumor Diagn. Ther. *3:* 96 (1982).
1012 Muhrer, K. H.; Burkhardt, M.; Bosslet, K.; Holzheimer, R.; Aigner, K. R.; Grebe, S.: Intraarterielle Infusion von monoklonalen Antikörpern bei Lebermetastasen colorektaler Karzinome. Int. Symp. Monoclonal Antibodies in Clinical Oncology, Homburg/Saar, 8.–9. 11. 1985.
1013 Mulshine, J. L.; Cuttitta, F.; Bibro, M.; Fedorko, J.; Fargion, S.; Little, C.; Carney, D. N.; Gazdar, A. F.; Minna, J. D.: Monoclonal antibodies that distinguish nonsmall cell from small cell lung cancer. J. Immun. *131:* 497 (1983).
1014 Murphy, W. H.; Herlocher, M. L.; Griep, J.: Immunization of C58 mice to line I_b leukemia. J. infect Dis. *112:* 28 (1963).
1015 Murray, G.: Experiments in immunity in cancer. Can. med. Ass. J. *79:* 249 (1958).
1016 Murray, D. R.; Cassel, W. A.; Torbin, A. H.; Olkowski, Z. L.; Moore, M. E.: Viral oncolysate in the management of malignant melanoma. II. Clinical studies. Cancer *40:* 680 (1977).
1017 Myers, W. L.; O'Keefe, M. L.: Use of capillary tube leukocyte adherence inhibition for the detection of tumor-specific immune response. Immunol. Commun. *10:* 407 (1981).
1018 Nadler, S. H.; Moore, G. E.: Immunotherapy of malignant disease. Archs Surg. *99:* 376 (1969).
1019 Nadler, L. M; Stashenko, P.; Hardy, R.; Kaplan, W. D.; Button, L. N.; Kufe, D. W.; Antman, K. H.; Schlossman, S. F.: Serotherapy of a patient with a monoclonal antibody directed against a human lymphoma-associated antigen. Cancer Res. *40:* 3147 (1980).
1020 Nadler, L. M.; Ritz, J.; Hardy, R.; Pesando, J. M.; Schlossman, S. F.: A unique cell surface antigen identifying lymphoid malignancies of B cell origin. J. clin. Invest. *67:* 134 (1981).
1021 Nagy, Z. A.; Ishii, N.; Baxevanis, C. N.; Klein, J.: Lack of Ir-gene control in T-cell responses restricted by allogeneic MHC molecules. Behring Inst. Mitt. *70:* 74 (1982).
1022 Nakano, T.; Imai, Y.; Sawada, J.; Osawa, T.: The use of various lectins for the separation of T-cell subsets; in Schauer, Glycoconjugates, Proc. of the 5th Int. Symp., Kiel, Sept. 1979, p. 452 (Thieme, Stuttgart 1979).
1023 Naor, D.: Suppressor cells: Permitters and promoters of malignancy? Adv. Cancer Res. *29:* 45 (1979).
1024 Naor, D.: Coexistence of immunogenic and suppressogenic epitopes in tumor cells and various types of macromolecules. Cancer Immunol. Immunother. *16:* 1 (1983).
1025 Natali, P. G.; Imai, K.; Wilson, B. S., Bigotti, A.; Cavaliere, R.; Pellegrino, M. A.; Ferrone, S.: Structural properties and tissue distribution of the antigens recognized by the monoclonal antibody 653.40S to human melanoma cells. J. natn. Cancer Inst. *67:* 591 (1981).
1026 Natali, P. G.; Wilson, B. S.; Imai, K.; Bigotti, A.; Ferrone, S.: Tisue distribution,

molecular profile, and shedding of a cytoplasmic antigen identified by the monoclonal antibody 465.12S to human melanoma cells. Cancer Res. *42:* 583 (1982).
1027 Nathan, C. F.: The release of hydrogen-peroxide from mononuclear phagocytes and its role in extracellular cytolysis; in van Furth, Mononuclear phagocytes, p. 1165 (Nijhoff, The Hague 1980).
1028 Nathan, C.; Cohn, Z.: Role of oxygen-dependent mechanisms in antibody-induced lysis of tumor cells by activated macrophages. J. exp. Med. *152:* 198 (1980).
1029 Nathrath, W. B. J.: Organ and tumour antigens in malignant disease: A review. J. R. Soc. Med. *71:* 755 (1978).
1030 Nayak, S. K.; Knotts, F. B.; Drogemuller, C. R.; Pilch, Y. H.: Detection of antibodies bound to tumor cell surface antigens with radioiodinated Staphylococcus aureus protein A (SPA). Cancer Immunol. Immunother. *5:* 243 (1979).
1031 Nelson, D. S.: Macrophages: Progress and problems. Clin. exp. Immunol. *45:* 225 (1981).
1032 Nelson, D. S.; Hopper, K. E.; Blanden, R. V.; Gardner, I. D.; Kearney, R.: Failure of immunogenic tumors to elicit cytolytic T cells in syngeneic hosts. Cancer Lett. *5:* 61 (1978).
1033 Nelson, M.; Nelson, D. S.: Macrophages and resistance to tumours. I. Inhibition of delayed-type hypersensitivity reactions by tumour cells and by soluble products affecting macrophages. Immunology *34:* 277 (1978).
1034 Nelson, M.; Nelson, D. S.: Macrophages and resistance to tumours. IV. The influence of age on the susceptibility of mice to the anti-inflammatory and anti-macrophage effects of tumour cell products. J. natn. Cancer Inst. *65:* 781 (1980).
1035 Nettleship, A.: Regression produced in the Murphy lymphosarcoma by the injection of heterologous antibodies. Am. J. Path. *21:* 527 (1945).
1036 Neville, M. E.: Human killer cells and natural killer cells: distinct subpopulations of Fc receptor-bearing lymphocytes. J. Immun. *125:* 2604 (1980).
1037 Newman, R. A.; Klein, P. J.; Uhlenbruck, G.; Citoler, P.; Karduck, D.: The presence and significance of the Thomsen-Friedenreich antigen in breast cancer. I. Serological studies. J. Cancer Res. clin. Oncol. *93:* 181 (1979).
1038 Ng, A.-K.; Pellegrino, M. A.; Imai, K.: HLA-A, B antigens, Ia-like antigens, and tumor-associated antigens on prostate carcinoma cell lines: Serologic and immunochemical analysis with monoclonal antibodies. J. Immun. *127:* 443 (1981).
1039 Ng, A.-K.; Giacomini, P.; Ferrone, S.: Monoclonal antibodies and immunologic approaches to malignant tumors. Adv. internal Med. *28:* 253 (1983).
1040 Nicolin, A.; Canti, G.; Marelli, O.; Veronese, F.; Goldin, A.: Chemotherapy and immunotherapy of L1210 leukemic mice with antigenic tumor sublines. Cancer Res. *41:* 1358 (1981).
1041 Nicolson, G. L.; Winkelhake, J. L.: Organ specificity of blood-borne metastasis determined by cell adhesion. Nature *255:* 230 (1975).
1042 Nicolson, G. L.; Poste, G.: The cancer cell: dynamic aspects and modification in cell-surface organization (first of two parts). New Engl. J. Med. *295:* 197 (1976).
1043 Nilsson, O.; Månsson, J.-E.; Brezicka, T.; Holmgren, J.; Lindholm, L.; Sörenson, S.; Yngvason, F.; Svennerholm, L.: Fucosyl-G_{M1} – a ganglioside associated with small cell lung carcinomas. Glycoconjugate J. *1:* 43 (1984).
1044 Nind, A. P. P.; Nairn, R. C.; Pihl, E.; Hughes, E. S. R.; Cuthbertson, A. M.; Rollo, A. J.: Autochthonous humoral and cellular immunoreactivity to colorectal carci-

noma: Prognostic significance in 400 patients. Cancer Immunol. Immunother. 7: 257 (1980).
1045 Ninnemann, J. L.: Melanoma-associated immunosuppression through B cell activation of suppressor T cells. J. Immun. *120:* 1573 (1978).
1046 Nishioka, K.; Irie, R. F.; Kawana, T.; Takeuchi, S.: Immunological studies on mouse mammary tumours. III. Surface antigens reacting with tumour-specific antibodies in immune adherence. Int. J. Cancer *4:* 139 (1969).
1047 Nordling, S. E.; Mayhew, E.: On the intracellular uptake of neuraminidase. Exp. Cell Res. *44:* 552 (1966).
1048 Nordquist, R.; Anglin, H.; Lerner, M. P.: Antibody-induced antigen redistribution and shedding from human breast cancer cells. Science *197:* 366 (1977).
1049 Normann, S. J.: Tumour cell threshold required for suppression of macrophage inflammation. J. natn. Cancer Inst. *60:* 1091 (1978).
1050 Normann, S. J.; Sorkin, E.: Cell-specific defect on monocyte function during tumour growth. J. natn. Cancer Inst. *57:* 135 (1976).
1051 North, R. J.: The concept of the activated macrophage. J. Immun. *121:* 806 (1978).
1052 North, R. J.; Kirstein, D. P.; Tuttle, R. L.: Subversion of host defence mechanisms by murine tumours. I. A circulating factor that suppresses macrophage-mediated resistance to infection. J. exp. Med. *143:* 559 (1976).
1053 North, R. J.; Kirstein, D. P.; Tuttle, R. L.: Subversion of host defence mechanisms by murine tumours. II. Counter influence of concomitant anti-tumour immunity. J. exp. Med. *143:* 574 (1976).
1054 North, R. J.; Spitalny, G. L.; Kirstein, D. P.: Antitumor defence mechanisms and their subversion; in Waters, Handbook of cancer immunology, vol. 2, p. 187 (Garland STPM Press, New York 1978).
1055 North, R. J.; Spitalny, G. L.; Berendt, M. J.: Significance of systemic macrophage activation in response to tumor growth; in van Furth, Mononuclear phagocytes, functional aspects, part II, p. 1655 (Nijhoff, The Hague 1980).
1056 Novogrodsky, A.; Katchalski, E.: Induction of lymphocyte transformation by periodate. FEBS Lett. *12:* 297 (1971).
1057 Nowinski, R.; Berglund, C.; Lane, J.; Lostrom, M.; Bernstein, J.; Young, W.; Hakamori, S.; Hill, L.; Cooney, M.: Human monoclonal antibody against Forssman antigen. Science *210:* 537 (1980).
1058 Nunn, M. E.; Herberman, R. B.; Holden, H. T.: Natural cell-mediated cytotoxicity in mice against non-lymphoid tumor cells and some normal cells. Int. J. Cancer *20:* 381 (1977).
1059 Nussenzweig, M. C.; Steinman, R. M.: Contribution of dendritic cells to stimulation of the murine syngeneic mixed leukocyte reaction. J. exp. Med. *151:* 1196 (1980).
1060 Nussenzweig, M. C.; Steinman, R. M.; Gutchinov, B.; Cohn, Z. A.: Dendritic cells are accessory cells for the generation of anti-TNP cytotoxic T cells. J. exp. Med. *152:* 1070 (1980).
1061 Nuti, M.; Teramoto, Y. A.; Mariani-Costantini, R.; Horan Hand, P.; Colcher, D.; Schlom, J.: A monoclonal antibody (B72.3) defines patterns of distribution of a novel tumor-associated antigen in human mammary carcinoma cell populations. Int. J. Cancer *29:* 539 (1982).
1062 Oberbarnscheidt, J.; Kölsch, E.: Direct blockade of antigen-reactive B lymphocytes

by immune complexes. An 'off' signal for precursors of IgM-producing cells provided by the linkage of antigen- and Fc-receptors. Immunology 35: 151 (1978).
1063 Oehler, J. R.; Lindsay, L. R.; Nunn, M. E.; Holden, H. T.; Herberman, R. B.: Natural cell-mediated cytotoxicity in rats. II. In vivo augmentation of NK-cell activity. Int. J. Cancer 21: 210 (1978).
1064 Örn, A.; Gidlund, M.; Ojo, E.; Grönvik, K.-O.; Anderson, J.; Wigzell, J.; Murgita, R. A.; Senik, A.; Gresser, I.: Factors controlling the augmentation of natural killer cells; in Herberman, Natural cell-mediated immunity against tumors, p. 581 (Academic Press, New York 1980).
1065 Örntoft, T. F.; Mors, N. P. O.; Eriksen, G.; Jacobsen, N. O., Poulsen, H. S.: Comparative immunoperoxidase demonstration of T-antigens in human colorectal carcinomas and morphologically abnormal mucosa. Cancer Res. 45: 447 (1985).
1066 Oettgen, H. F.: Tumor immunology. Behring Inst. Mitt. 63: 80 (1979).
1067 Oettgen, H. F.; Old, L. J.; McLean, E. P.; Carswell, E. A.: Delayed hypersensitivity and transplantation immunity by soluble antigens of chemically induced tumors in inbred guinea pigs. Nature 220: 295 (1968).
1068 Oettgen, H. F.; Hellstrom, K. E.: Principles of immunology: Tumor immmunology; in Holland, Frei, Cancer medicine, p. 1029 (Lea & Febiger, Philadelphia 1982).
1069 Okada, T. S.: Development of kidney-specific antigens: An immunohistological study. J. Embryol. exp. Morph. 13: 285 (1965).
1070 Okada, H.; Baba, T.: Rosette formation of human erythrocytes on cultured cells of tumour origin and activation of complement by cell membrane. Nature, Lond. 248: 521 (1974).
1071 Okuda, K.; Cullen, S. E.; Hilgers, J.; David, C.: Immune response-associated antigens on mouse leukemia cells. I. Detection of Ia antigens on GRSL cells. Transplantation 26: 153 (1978).
1072 Old, L. J.; Clarke, D. A.; Benacerraf, B.: Effect of bacillus Calmette-Guérin on transplanted tumors in the mouse. Nature 184: 291 (1959).
1073 Old, L. J.; Benacerraf, B.; Clarke, D. A.: The role of the reticuloendothelial system in the host reaction to neoplasia. Cancer Res. 21: 1281 (1961).
1074 Old, L. J.; Boyse, E. A.; Clarke, D. A.; Carswell, E. A.: Antigenic properties of chemically induced tumors. Ann. N. Y. Acad. Sci. 101: 80 (1962).
1075 Old, L. J.; Stockert, E.; Boyse, E. A.; Geering, G.: A study of passive immunization against a transplanted G+ leukemia with specific antiserum. Proc. Soc. exp. Biol. Med. 124: 63 (1967).
1076 Old, L. J.; Boyse, E. A.; Geering, G.; Oettgen, H. F.: Serologic approaches to the study of cancer in animals and in man. Cancer Res. 28: 1288 (1968).
1077 Old, L. J.; Stockert, E.; Boyse, E. A.; Kim, J. H.: Antigenic modulation: Loss of TL antigens from cells exposed to TL antibody. Study of the phenomenon in vitro. J. exp. Med. 127: 523 (1968).
1078 Old, L. J.: Tumor necrosis factor. Clinical Bulletin 6: 118 (1976).
1079 Oldham, R. K.; Djeu, J. K.; Cannon, G. B.; Siwarski, D.; Herberman, R. B.: Cellular microcytotoxicity in human tumor systems: Analysis of results. J. natn. Cancer Inst. 55: 1305 (1975).
1080 Oldham, R. K.; Fidler, I. J.; Talmadge, J. E.: Screening of biologicals and biological response modifiers. Akt. Onkol. 24: 93 (1985).

1081 Olstad, R.; Gaudernack, G.; Kaplan, G.; Seljelid, R.: T-and B-cell-independent activation of syngeneic macrophages by murine sarcoma cells. Cancer Res. 40: 2054 (1980).
1082 Olstad, R.; Kaplan, G.; Seljelid, R.: In vitro cytotoxicity of mouse macrophages activated by coculture with syngeneic sarcoma cells. Scand. J. Immunol. 16: 421 (1982).
1083 Olstad, R.; Kaplan, G.; Seljelid, R.: Tumour-activated macrophages as effector cells in a tumour neutralization assay in vivo. Acta pathol. microbiol. scand., C, Immunol. 91: 305 (1983).
1084 Omary, M. B.; Trowbridge, I. S.; Scheid, M. P.: T200 cell surface glycoprotein of the mouse. Polymorphism defined by the Ly-5 system of alloantigens. J. exp. Med. 151: 1311 (1980).
1085 Oppenheim, J. J.; Stadler, B. M.; Siraganian, R. P.; Mage, M.; Mathieson, B.: Lymphokines: Their role in lymphocyte responses. Properties of interleukin I. Fed. Proc. 41: 257 (1982).
1086 Order, S. E.; Donahue, V.; Knapp, R.: Immunotherapy of ovarian carcinoma. An experimental model. Cancer 32: 573 (1973).
1087 Orlando, R. A.; Craft, K.; Glick, M.; Wissler, R. W.: Specific enhancement of hepatoma growth using methylated bovine serum albumin and pertussis vaccine. Lab. Invest. 26: 735 (1972).
1088 Orlando, R. A.; Craft, K.; Glick, M.; Barbour, C.; Wissler, R. W.: Hepatoma growth. Enhancement after immunization with plasma membranes and adjuvant. Archs Path. 95: 229 (1973).
1089 Ortaldo, J. R.; Pestka, S.; Slease, R. B.; Rubenstein, M.; Herberman, R. B.: Augmentation of human K-cell activity. Scand. J. Immunol. 12: 365 (1980).
1090 Oth, D.; Berebbi, M.; Meyer, G.: Tumor-associated antigens in isoantigenic variants of a 3-methylcholanthrene-induced sarcoma. J. natn. Cancer Inst. 55: 903 (1975).
1091 Otu, A. A.; Russell, R. J.; Wilkinson, P. C.; White, R. G.: Alterations of mononuclear phagocyte function induces by Lewis lung carcinoma in C57Bl mice. Br. J. Cancer 36: 330 (1977).
1092 Ouchterlony, Ö.: In vitro method for testing the toxin-producing capacity of diphteria bacteria. Acta path. microbiol. scand. 25: 186 (1948).
1093 Ouchterlony, O.: Immunodiffusion and immunoelectrophoresis; in Weir, Handbook of experimental immunology, p. 655 (Blackwell, Oxford, Edinburgh 1967).
1094 Owen, L. N.; Bostock, D. E.; Lavelle, R. B.: Studies on therapy of osteosarcoma in dogs using BCG vaccine. J. Am. Vet. Radiol. Soc. 18: 27 (1977).
1095 Owen, L. N.: A comparative study of canine and human breast cancer. Invest. Cell Path. 2: 257 (1979).
1096 Paglieroni, T.; MacKenzie, M. R.: Studies on the pathogenesis of an immune defect in multiple myeloma. J. clin. Invest. 59: 1120 (1977).
1097 Panteleakis, P. N.; Larson, V. M.; Glenn, E. S.; Hilleman, M. R.: Prevention of viral and transplant tumors in hamsters employing killed and fragmented homologous tumor cell vaccines. Proc. Soc. exp. Biol. Med. 129: 50 (1968).
1098 Panteleakis, P. N.; Larson, V. M.; McAleer, W. J.; Hilleman, M. R.: Variations in immunogenicity of disrupted cells prepared from an adenovirus-7 hamster tumour cell line. J. natn. Cancer Inst. 46: 1195 (1971).

1099 Pape, G. R.; Troye, M.; Perlmann, P.: Characterization of cytolytic effector cells in peripheral blood of healthy individuals and cancer patients. I. Surface markers and K cell activity after separation of B cells and lymphocytes with Fc-receptors by column fractionation. J. Immunol. *118:* 1919 (1977).

1100 Paranjpe, M. S.; Boone, C. W.; Takeichi, N.: Specific paralysis of the anti-tumor cellular immune response produced by growing tumors studied with a radioisotope footpad assay. Ann. N. Y. Acad. Sci. *276:* 254 (1976).

1101 Parish, C. R.; Jackson, D. C.; McKenzie, I. F. C.: Low-molecular-weight Ia antigens in normal mouse serum. III. Isolation and partial chemical characterization. Immunogenetics *3:* 455 (1976).

1102 Parish, C. R.; Higgins, T. J.; McKenzie, I. F. C.: Comparison of antigens recognized by xenogeneic and allogeneic anti-Ia antibodies: Evidence for two classes of Ia antigens. Immunogenetics *6:* 343 (1978).

1103 Parish, C. R.; McKenzie, I. F. C.: A detailed serological analysis of a xenogeneic anti-Ia serum. Immunogenetics *6:* 183 (1978).

1104 Parish, C. R.; Higgins, T. J.; McKenzie, I. F. C.: Lymphocytes express Ia antigens of foreign haplotype following treatment with neuraminidase. Immunogenetics *12:* 1 (1981).

1105 Parker, J. W.; O'Brien, R. L.; Lukes, R. J.; Steiner, J.: Transformation of human lymphocytes by sodium periodate. Lancet *i:* 103 (1972).

1106 Parkinson, D. R.; Brightman, R. P.; Waksal, S. D.: Altered natural killer cell biology in C57Bl/6 mice after leukemogenic split-dose irradiation. J. Immun. *126:* 1460 (1981).

1107 Parmiani, G.: Histocompatibility antigens and tumour antigens. Cancer Immunol. Immunother. *8:* 215 (1980).

1108 Parmiani, G.; Meschini, A.; Invernizzi, G.; Carbone, G.: Tumor-associated transplantation antigens distinct from $H-2^k$-like antigens on a Balb/c ($H-2^d$) fibrosarcoma. J. natn. Cancer Inst. *61:* 1229 (1978).

1109 Parmiani, G.; Ballinari, D.: Genetic control of in vivo immunity to tumor-specific transplantation antigens of chemically induced murine fibrosarcomas. Int. J. Cancer *23:* 697 (1979).

1110 Parmiani, G.; Carbone, G.; Invernizzi, G.; Pierotti, M. A.; Sensi, M.; Rogers, M. J.; Appella, E.: Alien histocompatibility antigens on tumor cells. Immunogenetics *9:* 1 (1979).

1111 Parmiani, G.; Pierotti, M. A.: Generation of TSTA diversity. Looking for testable hypotheses. Cancer Immunol. Immunother. *14:* 133 (1983).

1112 Parodi, A. L.; Misdorp, W.; Mialot, J. P.; Mialot, M.; Hart, A. A. M.; Hurtrel, M.; Salomon, J. C.: Intratumoral BCG and Corynebacterium parvum therapy of canine mammary tumours before radical mastectomy. Cancer Immunol. Immunother. *15:* 172, (1983).

1113 Paswell, J. H.; Dayer, J. M.; Merler, E.: Increased prostaglandin production by human monocytes after membrane receptor activation. J. Immun. *123:* 115 (1979).

1114 Paucker, K.; Cantell, K.; Henle, W.: Quantitative studies on viral interference in suspended L cells. III. Effect of interfering viruses and interferon on the growth rate of cells. Virology *17:* 324 (1962).

1115 Pearson, G.; Freeman, G.: Evidence suggesting a relationship between polyoma

virus-induced transplantation antigen and normal embryonic antigen. Cancer Res. *28:* 1665 (1968).
1116 Pellis, N. R.; Yamagishi, H.; Shulan, D. J.; Kahan, B. D.: Use of preparative isoelectric focusing in Sephadex Gel Slab to separate immunizing and growth facilitating moieties in crude 3M KCl extracts of a murine fibrosarcoma. Cancer Immunol. Immunother. *11:* 53 (1981).
1117 Penn, I.: Tumor incidence in human allograft recipients. Transplant. Proc. *11:* 1047 (1979).
1118 Pennica, D.; Nedwin, G. E.; Hayflick, J. S.; Seeburg, P. H.; Derynck, R.; Palladino, M. A.; Kohr, W. J.; Aggarwal, B. B.; Goeddel, D. V.: Human tumour necrosis factor: precursor structure, expression and homology to lymphotoxin. Nature *312:* 724 (1984).
1119 Perlmann, P.; Perlmann, H.; Wigzell, H.: Lymphocyte mediated cytotoxicity in vitro induction and inhibition by humoral antibody and nature of effector cells. Transplant. Rev. *13:* 91 (1972).
1120 Perper, R. J.; Oronsky, A. L.; Sander, M.: The effects of BCG on extravascular mononucelar cell accumulation in vivo. Int. J. Cancer *17:* 670 (1976).
1121 Perry, L. L.; Benacerraf, B.; Greene, M. I.: Regulation of the immune response to tumor antigen. IV. Tumor antigen specific suppressor factor(s) bear I-J determinants and induce suppressor T cells in vivo. J. Immun. *121:* 2144 (1978).
1122 Perussia, B.; Trinchieri, G.; Cerottini, J.-C.: Analysis of human lymphocytes subpopulations responsible for antibody-dependent and spontaneous cell-mediated cytotoxicity. Transplant. Proc. *XI:* 793 (1979).
1123 Pesando, J. M.; Tomaselli, K. J.; Lazarus, H.; Schlossman, S. F.: Distribution and modulation of a human leukemia-associated antigen (CALLA). J. Immun. *131:* 2038 (1983).
1124 Petranyi, G.; Kiessling, R.; Povey, S.; Klein, G.; Herzenberg, L.; Wigzell, H.: The genetic control of natural killer cell activity and its association with in vivo resistance against a moloney lymphoma isograft. Immunogenetics *3:* 15 (1976).
1125 Pettinelli, C. B.; Ahmann, G. B.; Shearer, G. M.: Expression of both I-a and I-e/c subregion antigens on accessory cells required for in vitro generation of cytotoxic T lymphocytes against alloantigens or TNBS-modified syngeneic cells. J. Immun. *124:* 1911 (1980).
1126 Pickaver, A. H.; Ratcliffe, N. A.; Williams, A. E.; Smith, H.: Cytotoxic effects of peritoneal neutrophils on a syngeneic rat tumour. Nature new Biol. *235:* 186 (1972).
1127 Pierce, C. W.; Kapp, J. A.; Benacerraf, B.: Regulation by the H-2 gene complex of macrophage-lymphoid cell interactions in secondary antibody responses in vitro. J. exp. Med. *144:* 371 (1976).
1128 Pierce, C. W.; Aune, T. M.: Mechanism of action of soluble immune response suppressor (SIRS); in Hadden, Mullen, Advances in immunopharmacology, p. 397 (Pergamon, Oxford 1981).
1129 Piere, G. E.; DeVald, B. L.: Effects of human lymphocytes on cultured normal and malignant cells. Cancer Res. *35:* 1830 (1975).
1130 Pike, M. C.; Snyderman, R.: Depression of macrophage function by a factor produced by neoplasms: A mechanism for abrogation of immune surveillance. J. Immun. *117:* 1243 (1976).
1131 Pilch, Y. H.; Fritze, D.; Dekernion, J. B.; Ramming, K. P.; Kern, D. H.: Immuno-

therapy of cancer with immune RNA in animal models and cancer patients. Ann. N. Y. Acad. Sci. *277:* 592 (1976).

1132 Pinkuss, A.: Zur Vakzinationstherapie des Krebses. Berl. klin. Wschr. *50:* 1941 (1913).

1133 Plata, F.; MacDonald, H. R.; Sordat, B.: Studies on the distribution and origin of cytolytic T-lymphocytes present in mice-bearing Moloney murine sarcoma virus (MSV) induced tumors; in Clemmesen, Yohn, Comparative leukemia research 1975. Bibl. Haemat., vol. 43, p. 274 (Karger, Basel 1975).

1134 Playfair, J. H. L.: The role of antibody in T-cell response. Clin. exp. Immunol. *17:* 1 (1974).

1135 Plescia, O. J.; Smith, A. H.; Grinwich, K.: Subversion of immune system by tumor cells and role of prostaglandins. Proc. natn. Acad. Sci. USA *72:* 1848 (1975).

1136 Pollack, S. B.: Expression of oncofetal antigens on murine sarcomas characterized for expression of endogenous MuLV. Int. J. Cancer *22:* 344 (1978).

1137 Pollack, S. B.; Tam, M. R.; Nowinski, R. C.; Emmons, S. L.: Presence of T cell-associated surface antigens on murine NK cells. J. Immun. *123:* 1818 (1979).

1138 Polverini, P. J.; Cotran, R. S.; Gimbrone, M. A. Jr.; Unanue, E. R.: Activated macrophages induce vascular proliferation. Nature, Lond. *269:* 804 (1977).

1139 Porzsolt, F.; Tautz, C.; Schmidtberger, R.; Ax, W.: Zellelektrophoretische Untersuchungen zur Tumordiagnostik. Z. ImmunForsch. *147:* 352 (1974).

1140 Porzsolt, F.; Tautz, C.; Ax, W.: Electrophoretic mobility test. I. Modification to simplify the detection of malignant disease in man. Behring Inst. Mitt. *57:* 128 (1975).

1141 Poste, G.; Kirsh, R.; Fogler, W. E.; Fidler, I. J.: Activation of tumoricidal properties in mouse macrophages by lymphokines encapsulated in liposomes. Cancer Res. *39:* 881 (1979).

1142 Poste, G.; Doll, J.; Fidler, I. J.: Interactions among clonal subpopulations affect stability of the metastatic phenotype in polyclonal populations of B16 melanoma cells. Proc. natn. Acad. Sci. USA *78:* 6226 (1981).

1143 Poste, G.; Doll, J.; Brown, A. E.; Tzeng, J.; Zeidman, I.: Comparison of the metastatic properties of B16 melanoma clones isolated from cultured cell lines, subcutaneous tumors, and individual lung metastases. Cancer Res. *42:* 2770 (1982).

1144 Poston, R. N.; Morgan, R. S.: Interactions between soluble IgG, complement and cells in lymphocyte and monocyte ADCC. Immunology *50:* 461 (1983).

1145 Pottathil, R.; Huebner, R. J.; Meier, H.: Suppression of chemical (DEN) carcinogenesis in SWR/J mice by goat antibodies against endogenous murine leukemia viruses. Proc. Soc. exp. Biol. Med. *159:* 65 (1978).

1146 Pottathil, R., Meier, H., Huebner, R. J.: Suppression of leukemogenesis in hairless mice: by anti-type C viral immune gamma globulins. Naturwissenschaften *65:* 443 (1978).

1147 Poulter, L. W.: Antigen presenting cells in situ: their identification and involvement in immunopathology. Clin. exp. Immunol. *53:* 513 (1983).

1148 Poulter, L. W.; Seymour, G. J.; Duke, O.; Janossy, G.; Panayi, G.: Immunohistological analysis of delayed-type hypersensitivity in man. Cell. Immunol. *74:* 358 (1982).

1149 Powell, A. E.; Sloss, A. M.; Smith, R. N.; Makley, J. T.; Hubay, C. A.: Specific responsiveness of leukocytes to soluble extracts of human tumors. Int. J. Cancer *16:* 905 (1975).

1150 Powles, R. L.; Russel, J.; Lister, T. A.; Oliver, T.; Whitehouse, J. M. A.; Malpas, J.; Chapuis, B.; Crowther, D.; Alexander, P.: Immunotherapy for acute myelogenous leukemia: A controlled clinical study 2½ years after entry of the last patient. Br. J. Cancer *35:* 265 (1977).

1151 Prager, M. D.: Immune response to modified tumor cells; in Walborg, 28th Annual Symp. on Fundamental Cancer Res. of the U. of Texas System Cancer Center, M. D. Anderson Hospital and Tumor Institute, p. 523 (Williams & Wilkins, Baltimore 1975).

1152 Prager, M. D.: Specific cancer immunotherapy. Cancer Immunol. Immunother. *3:* 157 (1978).

1153 Prager, M. D.; Derr, I.; Swann, A.; Cotropia, J.: Immunization with chemically modified lymphoma cells. Cancer Res. *31:* 1488 (1971).

1154 Prager, M. D.; Hollinshead, A. C.; Ribble, R. J.; Derr, I.: Induction of immunity to a mouse lymphoma by multiple methods, including vaccination with soluble membrane fractions. J. natn. Cancer Inst. *51:* 1603 (1973).

1155 Prager, M. D.; Ticaric, S.; Merrill, C. L.: Tumor host relationship in immune response to modified lymphoma cells. Proc. Am. Ass. Cancer Res. *13:* 103 (1972).

1156 Prager, M. D.; Baechtel, F. S.: Methods for modification of cancer cells to enhance their antigenicity; in Busch, Methods in cancer research, vol. 9, p. 339 (Academic Press, New York 1973).

1157 Prager, M. D.; Hollinshead, A. C.; Ribble, R. J.; Derr, I.: Induction of immunity to a mouse lymphoma by multiple methods, including vaccination with soluble membrane fractions. J. natn. Cancer Inst. *51:* 1603 (1972).

1158 Prager, M. D.; Mehta, J. M.: Enzymes as immunosuppressants: Basic considerations. Transplant. Proc. *5:* 1171 (1973).

1159 Prager, M. D.; Baechtel, F. S.; Ribble, R. J.; Ludden, C. M.; Mehta, J. M.: Immunological stimulation with modified lymphoma cells in a minimally responsive tumor-host system. Cancer Res. *34:* 3203 (1974).

1160 Prager, M. D.; Ludden, C. M.; Mandy, W. J.; Allison, J. P.; Kitto, G. B.: Endotoxin stimulated immune response to modified lymphoma cells. J. natn. Cancer Inst. *54:* 773 (1975).

1161 Prager, M. D.; Gordon, W. C.; Baechtel, F. S.: Immunogenicity of modified tumour cells in syngeneic hosts. Ann. N. Y. Acad. Sci. *276:* 61 (1976).

1162 Prehn, R. T.: The immune reaction as a stimulator of tumor growth. Science *176:* 170 (1972).

1163 Prehn, R. T.: Tumor progression and homeostasis. Adv. Cancer Res. *23:* 203 (1976).

1164 Prehn, R. T.: Immunostimulation of the lymphodependent phase of neoplastic growth. J. natn. Cancer Inst. *59:* 1043 (1977).

1165 Prehn, R. T.: Review/Commentary: The dose-response curve in tumor-immunity. Int. J. Immunopharm. *5:* 255 (1983).

1166 Prehn, R. T.; Main, J. M.: Immunity to methylcholanthrene-induced sarcoma. J. natn. Cancer Inst. *18:* 759 (1957).

1167 Prehn, R. T.; Lappé, M. A.: An immunostimulation theory of tumor development. Transplant. Rev. *7:* 26 (1971).

1168 Pressman, D.; Sternberger, L. A.: The relative rates of iodination of serum components and the effect of iodination on antibody activity. J. Am. chem. Soc. *22:* 2226 (1950).
1169 Price, M. R.; Dennick, R. G.; Robins, R. A.; Baldwin, R. W.: Modification of the immunogenicity and antigenicity of rat hepatoma cells. I. Cell-surface stabilization with glutaraldehyde. Br. J. Cancer *39:* 621 (1979).
1170 Procter, J.; Rudenstam, C. M.; Alexander, P.: Increased incidence of lung metastases following treatment of rats bearing hepatomas with irradiated tumour cells and the beneficial effect of C. parvum in this system. Biomedicine *19:* 248 (1973).
1171 Proctor, J. W.; Auclair, B. G.; Stokowski, L.: Endocrine factors and the growth and spread of B16 melanoma. J. natn. Cancer Inst. *57:* 1197 (1976).
1172 Pross, H. F.; Baines, M. G.: Spontaneous human lymphocyte-mediated cytotoxicity against tumor target cells. Cancer Immunol. Immunother. *3:* 75 (1977).
1173 Pruss, R. M.; Mirsky, R.; Raff, M. C.; Thorpe, R.; Dowding, A. J.; Anderton, B. H.: All classes of intermediate filaments share a common antigenic determinant defined by a monoclonal antibody. Cell *27:* 419 (1981).
1174 Ptak, W.; Naidorf, K. F.; Gershon, R. K.: Interference with the transmission of T cell-derived messages by macrophage membranes. J. Immun. *119:* 444 (1977).
1175 Ptak, W.; Zembala, M.; Gershon, R. K.: Intermediary role of macrophages in the passage of suppressor signals between T-cell subsets. J. exp. Med. *148:* 424 (1978).
1176 Ptak, W.; Rozycka, D.; Askenase, P. W.; Gershon, R. K.: Role of antigen-presenting cells in the development and persistence of contact hypersensitivity. J. exp. Med. *151:* 362 (1980).
1177 Pukel, C. S.; Lloyd, K. O.; Travassos, L. R.; Dippold, W. G.; Oettgen, H. F.; Old, L. J.: G_{D3}, a prominent ganglioside of human melanoma. J. exp. Med. *155:* 1133 (1982).
1178 Rabinovitch, M.; Hamburg, S. I.: Macrophage activation by interferon or interferon inducers; in van Furth, Mononuclear phagocytes, functional aspects, part II, p. 59 (Nijhoff, The Hague 1980).
1179 Radov, L. A.; Haskill, J. S.; Korn, J. H.: Host immune potentiation of drug responses to a murine mammary adenocarcinoma. Int. J. Cancer *17:* 773 (1976).
1180 Raftery, N. J.; Poulter, L. W.; Janossy, G.; Sweny, P.; Fernando, O.; Moorhead, J.: Heterogeneity of HLA-DR + ve cells in normal human kidney. Immunohistological and cytochemical characterization of discrete cell populations. J. clin. Path. *36:* 734 (1983).
1181 Rager-Zisman, B.; Grose, C.; Bloom, B. R.: Mechanism of selective nonspecific cell-mediated cytotoxicity of virus-infected cells. Nature, Lond. *260:* 369 (1976).
1182 Rahman, A. F. R.; Longenecker, B. M.: A monoclonal antibody specific for the Thomsen-Friedenreich cryptic T antigen. J. Immun. *129:* 2021 (1982).
1183 Rainer, H.; Kovats, E.; Lehmann, H. G.; Micksche, M.; Rauhs, R.; Sedlacek, H. H.; Seidl, W.; Schemper, M.; Schiessel, R.; Schweiger, B.; Wunderlich, M.: Effectiveness of postoperative adjuvant therapy with cytotoxic chemotherapy (cytosine arabinoside, mitomycin C, 5-fluorouracil) or immunotherapy (neuraminidase-modified allogeneic cells) in the prevention or recurrence of Duke's B and C colon cancer. Recent Results in Cancer Res. *79:* 42 (1981).
1184 Rainer, H.; Dittrich, C.; Rauhs, R.; Schemper, M.; Schiessel, R.; Wunderlich, M.; Micksche, M.; Vetterleich, M.; Kovats, E.; Lehmann, H. G.; Sedlacek, H. H.: Effec-

tiveness of postoperative adjuvant therapy with cytotoxic chemotherapy (cytosine arabinoside, mitomycin C, 5-fluorouracil) of immunotherapy (neuraminidase-modified allogeneic cells) in the prevention or recurrence of Duke's B and C colon cancer. Proc. Am. Cancer Ass. *25:* 479 (1984).

1185 Rajewsky, K.; Schirrmacher, V.; Nase, S.; Jerne, N. K.: The requirement of more than one antigenic determinant for immunogenicity. J. exp. Med. *129:* 1131 (1969).

1186 Ramming, K. P.; Pilch, Y. H.: Transfer of tumor-specific immunity with RNA; inhibition of growth of murine tumor isografts. J. natn. Cancer Inst. *46:* 735 (1971).

1187 Ransom, J. H.; Evans, C. H.; DiPaolo, J. A.: Lymphotoxin prevention of diethylnitrosamine carcinogenesis in vivo. J. natn. Cancer Inst. *69:* 741 (1982).

1188 Rao, G. R.; Rawls, W. E.; Perey, D. Y. E.; Tompkins, W. A. F.: Macrophage activation in congenitally athymic mice raised under conventional or germ-free conditions. J. reticuloendoth. Soc. *21:* 13 (1977).

1189 Rao, V. S.; Bennett, J. A.; Shen, F. W.; Gershon, R. K.; Mitchell, M. S.: Antigen-antibody complexes generate Lyt 1 inducers of suppressor cells. J. Immun. *125:* 63 (1980).

1190 Rapp, F.; Butel, J. S.; Melnick, J. L.: Virus-induced intranuclear antigen in cells transformed by papova-virus SV40. Proc. Soc. exp. Biol. Med. *116:* 1131 (1964).

1191 Rapp, F.; O'Connor, T. E.: Virology; in Holland, Frei, Cancer medicine, p. 12 (Lea & Febiger, Philadelphia 1982).

1192 Reinherz, E. L.; Schlossman, S. F.: The differentiation and function of human T lymphocytes. Cell *19:* 821 (1980).

1193 Reinisch, C. L.; Andrew, S. L.; Schlossman, S. F.: Suppressor cell regulation of immune response to tumors: Abrogation by adult thymectomy. Proc. natn. Acad. Sci. *74:* 2989 (1977).

1194 Reizenstein, P.; Mathé, G.: Biological response modifiers and differentiation inducers: Current status in cancer research. Drugs *26:* 185 (1983).

1195 Remacle-Bonnet, M. M.; Pommier, G. J.; Kaplanski, S.: Inhibition of normal allogeneic lymphocyte mitogenesis by a soluble inhibitor extracted from human colonic carcinoma. J. Immun. *117:* 1145 (1976).

1196 Renoux, G.: Biological augmenting agents; in Sirois, Rola-Pleszczynski, Immunopharmacology, p. 287 (Elsevier, Amsterdam 1982).

1197 Reynolds, C. W.; Herberman, R. B.: In vitro augmentation of rat natural killer (NK) cell activity. J. Immun. *126:* 1581 (1981).

1198 Reynolds, C. W.; Timonen, T.; Herberman, R. B.: Natural killer (NK) cell activity in the rat. I. Isolation and characterization of the effector cells. J. Immun. *127:* 282 (1981).

1199 Rhodes, J.: Receptor for monomeric IgM on guinea pig splenic macrophages. Nature, Lond. *243:* 527 (1973).

1200 Rhodes, J.: Altered expression of human monocyte Fc receptors in malignant disease. Nature, Lond. *265:* 253 (1977).

1201 Rhodes, J.: Resistance of tumour cells to macrophages. A short review. Cancer Immunol. Immunother. *7:* 211 (1980).

1202 Rhodes, J.; Bishop, M.; Benfield, J.: Tumor surveillance: how tumors may resist macrophage mediated host defense. Science *203:* 179 (1979).

1203 Riesenfeld, I.; Örn, A.; Gidlund, M.; Axberg, I.; Alm, G. V.; Wigzell, H.: Positive

correlation between in vitro NK activity and in vivo resistance towards AKR lymphoma cells. Int. J. Cancer 25: 399 (1980).
1204 Rigby, P. G.: Prolongation of survival of tumor-bearing animals by transfer of «Immune» RNA with DEAE dextran. Nature, Lond. 221: 968 (1969).
1205 Rios, A.; Simmons, R. L.: Active specific immunotherapy of minimal residual tumor: Excision plus neuraminidase-treated tumor cells. Int. J. Cancer 13: 71 (1974).
1206 Rios, A.; Simmons, R. L.: Comparative tumor immunoregressive effect of neuraminidase concavalin A or irradiated tumor cells. Fed. Proc. 33: 615 (1974).
1207 Risley, E. H.: The gilman-coca vaccine emulsion treatment of cancer. Boston Med. Surg. J. 165: 784 (1911).
1208 Ritz, J.; Pesando, J. M.; Sallan, S. E.; Clavell, L. A.; Notis-McConarty, J.; Rosenthal, P.; Schlossman, S. F.: Serotherapy of acute lymphoblastic leukemia with monoclonal antibody. Blood 58: 141 (1981).
1209 Robert-Guroff, M.; Gallo, R. C.: Establishment of an etiologic relationship between the human T-cell leukemia/lymphoma virus (HTLV) and adult T-cell leukemia. Blut 47: 1 (1983).
1210 Rockoff, S. D.; McIntire, K. R.; Ng, A. K.; Princler, G. L.; Herberman, R. B.; Larson, J. N.: Sensitive and convenient quantitation of antibody binding to cellular antigens using glutaraldehyde preserved cells. J. immunol. Method 26: 369 (1979).
1211 Roder, J. C.: Duwe, A.: The beige mutation in the mouse selectively impairs natural killer cell function. Nature, Lond. 278: 451 (1979).
1212 Roder, J. C.; Haliotis, T.: Do NK cells play a role in antitumor surveillance? Immunology today 1: 96 (1980).
1213 Roder, J. C.; Haliotis, T.; Klein, M.; Korec, S.; Jett, J. R.; Ortaldo, J.; Herberman, R. B.; Katz, P.; Fauci, A. S.: A new immunodeficiency disorder in humans involving NK cells. Nature 284: 553 (1980).
1214 Rodrigues-Netto, N.; Napoli, F.; Caserta Lemos, G.: Supporting action of chemotherapy and immunotherapy in carcinoma of the bladder. Rev. Paul. Med. 94: 51 (1979).
1215 Rogers, M. J.; Appella, E.; Pierotti, M. A.; Invernizzi, G.; Parmiani, G.: Biochemical characterization of alien H-2 antigen expressed on a methylcholanthrene-induced tumour. Proc. natn. Acad. Sci. USA 76: 1415 (1979).
1216 Rogers, M. J.; Law, L. W.; Pierotti, M. A.; Parmiani, G.: Separation of the tumor-associated transplantation antigen (TATA) from the alien H-2^k antigens expressed on a methylcholanthrene-induced tumor. Int. J. Cancer 25: 105 (1980).
1217 Rojan-Grgas, J.; Milas, L.: Effect of tumour cell culture media and sera from tumour hosts on spreading phagocytosis and antitumour cytotoxicity of C. parvum-activated murine macrophages. Cancer Immunol. Immunother. 6: 169 (1979).
1218 Rosenau, W.: Target cell destruction. Fed. Proc. 27: 34 (1968).
1219 Rosenau, W.: Lymphotoxin properties, role and mode of action. Int. J. Immunopharmacol. 3: 1 (1981).
1220 Rosenberg, E. B.; Herberman, R. B.; Levine, P. H.; Halterman, R. H.; McCoy, J. L.; Wunderlich, J. R.: Lymphocyte cytotoxicity reactions to leukemia-associated antigens in identical twins. Int. J. Cancer 9: 648 (1972).
1221 Rosenberg, S. A.; Henrichon, M.; Coyne, J. A.; David, J. R.: Guinea pig lymphotoxin (LT). I. In vitro studies of LT produced in response to antigen stimulation of lymphocytes. J. Immunol. 110: 1623 (1973).

1222 Rosenthal, A. S.: Determinant selection and macrophage function in genetic control of the immune response. Immunol. Rev. *40:* 135 (1978).
1223 Rosenthal, A. S.: Regulation of the immune response – role of the macrophage. New Engl. J. Med. *303:* 1153 (1980).
1224 Rosenthal, S. R.: Cancer control by stimulation of the immune system. Bull. Inst. Pasteur *81:* 55 (1983).
1225 Rosenthal, A. S.; Shevach, E. M.: The function of macrophages in antigen recognition by guinea pig T lymphocytes. I. Requirement for histocompatible macrophages and lymphocytes. J. exp. Med. *138:* 1194 (1973).
1226 Rosenthal, A. S.; Barcinski, M. A.; Rosenwasser, L. J.: Function of macrophages in genetic control of immune responsiveness. Fed. Proc. *37:* 79 (1978).
1227 Ross, M. W.; Tiangco, G. J.; Horn, P.; Hiserodt, J. C.; Granger, G. A.: The LT system in experimental animals. III. Physicochemical characteristics and relationships of lymphotoxin (LT) molecules released in vitro by activated lymphoid cells from several animal species. J. Immunol. *123:* 325 (1979).
1228 Rossi, F.; Romeo, D.; Patriarca, P.: Mechanism of phagocytosis-associated oxidative metabolism in polymorphonuclear leukocytes and macrophages. J. reticuloendoth. Soc. *12:* 127 (1972).
1229 Rothauge, C. F.; Kraushaar, J.; Gutschank, S.: Spezifische immunologische Intensivtherapie des entgleisten metastasierenden Prostata-Karzinoms. Diagnostik & Intensivtherapie *1:* 1 (1981).
1230 Rothauge, C. F.; Kraushaar, J.: Immuntherapie metastasierender Prostata- und hypernephroider Nierenkarzinome. Diagnostik *17:* 22 (1984).
1231 Rouzer, C. A.; Scott, W. A.; Hamill, A. L.; Cohn, Z. A.: Dynamics of leukotriene C production by macrophages. J. exp. Med. *152:* 1236 (1980).
1232 Rowland, G. F.: Use of antibodies to target drugs to tumour cells. Clin. Immunol. Allergy *3:* 235 (1983).
1233 Rowland, G. F.; Edwards, A. J.; Sumner, M. R.: Thymic dependency of tumor-induced immunodepression. J. natn. Cancer Inst. *50:* 1329 (1973).
1234 Rubin, R. H.; Cosimi, A. B.; Goetzl, E. J.: Defective human mononuclear leukocyte chemotaxis as an index of host resistance to malignant melanoma. Clin. Immunol. Immunopathol. *6:* 376 (1976).
1235 Rubin, B. Y.; Anderson, S. L.; Sullivan, S. A.; Williamson, B. D.; Carswell, E. A.; Old, L. J.: High affinity binding of ^{125}I-labeled human tumor necrosis factor (LuKII) to specific cell surface receptors. J. exp. Med. *162:* 1099 (1985).
1236 Rühl, H.; Fülle, H. H.; Kopeen, K. M.; Schwerdtfeger, R.: Adjuvant specific immunotherapy in maintenance treatment of adult acute non-lymphocytic leukaemia. Klin. Wschr. *59:* 1189 (1981).
1237 Ruff, M. R.; Gifford, G. E.: Tumor necrosis factor; in Pick, Lymphokines 2, p. 235 (Academic Press, New York 1981).
1238 Rundell, J. O.; Evans, C. H.: Biological characterization of guinea pig lymphotoxin production. Immunopharmacol. *2:* 19 (1979).
1239 Rundell, J. O.; Evans, C. H.: Species specificity of guinea pig and human lymphotoxin colony inhibitory activity. Immunopharmacology *3:* 9 (1981).
1240 Ruscetti, F. W.; Gallo, R. C.: Human T-lymphocyte growth factor: Regulation of growth and function of T lymphocytes. Blood *57:* 379 (1981).
1241 Russell, S. W.; Doe, W. F.; Hoskins, R. G.; Cochrane, C. G.: Inflammatory cells in

solid murine neoplasms. I. Tumor disaggregation and identification of constituent inflammatory cells. Int. J. Cancer *18:* 322 (1976).

1242 Russell, S. W.; McIntosh, A. T.: Macrophages isolated from regressing Moloney sarcomas are more cytotoxic than those recovered from progressing sarcomas. Nature, Lond. *268:* 69 (1977).

1243 Russel, J. H.; Ginns, L. C.; Terres, G.; Eisen, H. N.: Tumor antigens as inappropriately expressed normal alloantigens. J. Immunol. *122:* 912 (1979).

1244 Rygaard, J.; Povlsen, C. O.: The mouse mutant nude does not develop spontaneous tumours. An argument against immunological surveillance. Acta pathol. microbiol. scand., B *82:* 99 (1974).

1245 Ryan, G. B.: Acute inflammation. Am. J. Path. *86:* 184 (1977).

1246 Ryser, J. E.; Sordat, B.; Cerottini, J.-C.; Brunner, K. T.: Mechanism of target lysis by cytotoxic T lymphocytes. I. Characterization of specific lymphocytes-target cell conjugates separated by velocity sedimentation. Eur. J. Immunol. *7:* 110 (1977).

1247 Sachs, D. H.; Cone, J. L.: A mouse B-cell alloantigen determined by gene(s) linked to the major histocompatibility complex. J. exp. Med. *138:* 1289 (1973).

1248 Sadler, T. E.; Castro, J. E.: Lack of immunological and antitumor effects of orally administered C. parvum in mice. Br. J. Cancer *31:* 359 (1975).

1249 Sadler, T. E.; Castro, J. E.: Effect of C. parvum and surgery on the Lewis lung carcinoma and its metastases. Br. J. Surg. *63:* 292 (1976).

1250 Saksela, E.; Timonen, T.; Ranki, A.; Häyry, P.: Morphological and functional characterization of isolated effector cells responsible for human natural killer activity to fetal fibroblasts and to cultured cell line targets. Immunol. Rev. *44:* 71 (1979).

1251 Saksela, E.; Timonen, T.: Morphology and surface properties of human NK cells; in Herberman, Natural cell-mediated immunity against tumors, p. 173 (Academic Press, New York 1980).

1252 Saksela, E.; Timonen, T.; Virtanen, I.; Cantell, K.: Regulation of human natural killer activity by interferon; in Herberman, Natural cell-mediated immunity against tumors, p. 645 (Academic Press, New York 1980).

1253 Salaman, M. R.: The state of transfer factor. Immunology today *3:* 4 (1982).

1254 Salvin, S. B.; Youngner, J. S.; Nishio, J.; Neta, R.: Tumor suppression by a lymphokine released into the circulation of mice with delayed hypersensitivity. J. natn. Cancer Inst. *55:* 1233 (1975).

1255 Sanderson, C. J.: The mechanism of T cell mediated cytotoxicity. V. Morphological studies by electron microscopy. Proc. R. Soc. Lond. B *198:* 315 (1977).

1256 Sanderson, C. J.; Frost, P.: The induction of tumour immunity in mice using glutaraldehyde-treated tumour cells. Nature, Lond. *248:* 690 (1974).

1257 Santoh, D.; Trinchieri, G.; Zmijewski, C. M.; Koprowski, H.: HLA-related control of spontaneous and antibody-dependent cell-mediated cytotoxic activity in humans. J. Immunol. *117:* 765 (1976).

1258 Santoli, D.; Trichieri, G.; Moretta, L.; Zmijewski, C. M.; Koprowski, H.: Spontaneous cell-mediated cytotoxicity in humans. Distribution and characterization of the effector cells. Clin. exp. Immunol. *33:* 309 (1978).

1259 Santoli, D.; Koprowski, H.: Mechanisms of activation of human natural killer cells against tumor and virus-infected cells. Immunol. Rev. *44:* 125 (1979).

1260 Santoni, A.; Riccardi, C.; Barlozzari, T.; Herberman, R. B.: Suppression of activity of mouse natural killer (NK) cells by activated macrophages from mice treated with pyran copolymer. Int. J. Cancer 26: 837 (1980).

1261 Savary, C. A.; Lotzova, E.: Suppression of natural killer cell cytotoxicity by splenocytes from Corynebacterium parvum-injected, bone marrow-tolerant, and infant mice. J. Immun. *120:* 239 (1978).

1262 Sawada, J.; Shioiri-Nakano, K.; Osawa, T.: Cytotoxic activity of purified guinea pig lymphotoxin against various cell lines. Jap. J. exp. Med. *46:* 263 (1976).

1263 Sawada, J.; Kobayashi, Y.; Osawa, T.: The effect of pulse treatment of target cells with guinea pig lymphotoxin and the nature of its binding to target cells. Jap. J. exp. Med. *47:* 93 (1977).

1264 Sbarra, A. J.; Karnovsky, M. L.: The biochemical basis of phagocytosis. I. Metabolic changes during the ingestion of particles by polymorphonuclear leukocytes. J. biol. Chem. *234:* 1355 (1959).

1265 Schäfer, W.; Schwarz, H.; Thiel, H.-J.; Wecker, E.; Bolognesi, D. P.: Properties of mouse leukemia viruses. III. Serum therapy of virus-induced murine leukemias. Virology *75:* 401 (1976).

1266 Schechter, B.; Treves, A. J.; Feldman, M.: Specific cytotoxicity in vitro of lymphocytes sensitized in culture against tumor cells. J. natn. Cancer Inst. *56:* 975 (1976).

1267 Schechter, B.; Feldman, M.: Suppressor cells prevent host resistance to tumor growth. Naturwissenschaften *66:* 140 (1979).

1268 Scheidegger, J. J.: Une micro-méthode de l'immuno-électrophorèse. Int. Archs Allergy *7:* 103 (1955).

1269 Scheinberg, L. C.; Levine, M. C.; Suzuki, K.; Terry, R. D.: Induced host resistance to a transplantable mouse glioma. Cancer Res. *22:* 67 (1962).

1270 Schick, H.-J.; Schmidtberger, R.: Affinity of Vibrio cholerae neuraminidase to different human sialoglycoproteins. Behring Inst. Mitt. *55:* 123 (1974).

1271 Schick, H. J.; Zilg, H.: Production and quality control of therapeutically applicable Vibrio cholerae neuraminidase (VCN); in Griffith, Regamey, Dev. biol. Standard., vol. 38, p. 81 (Karger, Basel 1978).

1272 Schirrmacher, V.; Fogel, M.; Russmann, E.; Bosslet, K.; Altevogt, P.; Beck, L.: Antigenic variation in cancer metastasis: Immune escape versus immune control. Cancer Metastasis Reviews *1:* 241 (1982).

1273 Schlager, S. I.; Paque, R. E.; Dray, S.: Complete and apparently specific local tumor regression using syngeneic or xenogeneic «tumor immune» RNA extract. Cancer Res. *35:* 1907 (1975).

1274 Schlipköter, H.-W.; Stiller-Winkler, R.; Idel, H.; Knopp, J.: Semiquantitative Bestimmung von Antikörpern gegen tumorassoziierte Antigene in Seren von Bronchialkarzinomträgern mit Hilfe der Anti-Komplement-Immunfluoreszenz. Zentbl. Bakt. Hyg., I. Abt. Orig. B *174:* 105 (1981).

1275 Schlom, J.; Wunderlich, D.; Teramoto, Y. A.: Generation of human monoclonal antibodies reactive with human mammary carcinoma cells. Proc. natn. Acad. Sci. USA *77:* 6841 (1980).

1276 Schlom, J.; Greiner, J.; Hand, P. H.; Colcher, D.; Inghirami, G.; Weeks, M.; Pestka, S.; Fisher, P. B.; Noguchi, P.; Kufe, D.: Monoclonal antibodies to breast cancer-associated antigens as potential reagents in the management of breast cancer. Cancer *54:* 2777 (1984).

1277 Schmähl, D.; Habs, M.; Lorenz, M.; Wagner, I.: Occurrence of second tumors in man after anticancer drug treatment. Cancer Treat. Rev. *9:* 167 (1982).

1278 Schmidt, W.; Leben, L.; Atfield, G.; Festenstein, H.: Variations of expression of histocompatibility antigens on tumor cells: absence of H-2K^k gene products from a Gross-virus-induced leukemia in Balb.K. Immunogenetics *14:* 323 (1981).

1279 Schmidt, W.; Festenstein, H.: Resistance to cell-mediated cytotoxicity is correlated with reduction of H-2K gene products in AKR leukemia. Immunogenetics *16:* 257 (1982).

1280 Schnegg, J. F.; Diserens, A. C.; Carrel, S.; Accolla, R. S.; de Tribolet, N.: Human glioma-associated antigens detected by monoclonal antibodies. Cancer Res. *41:* 1209 (1981).

1281 Schorlemmer, H. U.; Allison, A. C.: Effects of activated complement components on enzyme secretion by macrophages. Immunology *31:* 781 (1976).

1282 Schorlemmer, H. U.; Sedlacek, H. H.; Seiler, F. R.; Bitter-Suermann, D.: Endogenous C3 as a trigger for macrophage stimulation. In: Proc. 4th Intl. Congress of Immunology, Paris (1980).

1283 Schorlemmer, H. U.; Sedlacek, H. H.; Seiler, F. R.; Bitter-Suermann, D.: Is the activation of macrophages by immune complexes (F(ab$_2$)IgG) dependent on endogenous C3? In: Proc.: 11. Arbeitstagung über Leukozytenkulturen, Erlangen, 6.–7. März (1980).

1284 Schorlemmer, H. U.; Bosslet, K.; Sedlacek, H. H.: Activation of macrophages for killing of tumor cells by the immunomodulator bestatin. Int. J. Immunopharmacol. *4:* 4 (1982).

1285 Schorlemmer, H. U.; Bosslet, K.; Sedlacek, H. H.: Tumoricidal activity of mononuclear phagocytes induced by the immunomodulating dipeptide bestatin. Immunobiology *162:* 419 (1982).

1286 Schorlemmer, H. U.; Lüben, G. H. A.; Sedlacek, H. H.: Activation of enhanced mononuclear phagocyte functions by the immunomodulator bestatin. Immunobiology *162:* 419 (1982).

1287 Schorlemmer, H. U.; Bosslet, K.; Sedlacek, H. H.: Ability of the immunomodulating dipeptide bestatin to activate cytotoxic mononuclear phagocytes. Cancer Res. *43:* 4148 (1983).

1288 Schorlemmer, H. U.; Bosslet, K.; Dickneite, G.; Lüben, G.; Sedlacek, H. H.: Studies on the mechanisms of action of the immunomodulator bestatin in various screening test systems. Behring Inst. Mitt. *74:* 157 (1984).

1289 Schroit, A. J.; Fidler, I. J.: Effects of liposome structure and lipid composition on the activation of the tumoricidal properties of macrophages by liposomes containing muramyl dipeptide. Cancer Res. *42:* 161 (1982).

1290 Schuepbach, J.; Sauter, C.: Inverse correlation of antiviral antibody titers and the remission length in patients treated with viral oncolysate: A possible new prognostic sign in acute myelogenous leukemia. Cancer *48:* 1363 (1981).

1291 Schuepbach, J.; Kalyanaraman, V. S.; Sarngadharan, M. G.; Blattner, W. A.; Gallo, R. C.: Antibodies against three purified proteins of the human type C retrovirus, human T-cell leukemia-lymphoma virus, in adult T-cell leukemia-lymphoma patients and healthy blacks from the Caribbean. Cancer Res. *43:* 886 (1983).

1292 Schultz, G. S.; Ebner, K. E.: Measurement of α-lactalbumin in serum and mammary tumors of rats by radioimmunoassays. Cancer Res. *37:* 4482 (1977).

1293 Schultz, R. M.; Papametheakis, J. D.; Chirigos, M. A.: Direct activation in vitro of mouse peritoneal macrophages by pyran copolymer (NSC 46015). Cell. Immunol. *29:* 403 (1977).

1294 Schultz, R. M.; Papamatheakis, J. D.; Chirigos, M. A.: Interferon: An inducer of macrophage activation by polyanions. Science *197:* 674 (1977).

1295 Schultz, R. M.; Chirigos, M. A.: Similarities among factors that render macrophages tumoricidal in lymphokine and interferon preparations. Cancer Res. *38:* 1003 (1978).

1296 Schultz, R. M.; Chirigos, M. A.; Heine, U. I.: Functional and morphological characteristics of interferon-treated macrophages. Cell. Immunol. *35:* 84 (1978).

1297 Schultz, R. M.; Chirigos, M. A.; Ravlidis, N. A.; Younger, J. S.: Macrophage activation and antitumor activity of Brucella abortus ether extract, Bru-Pel. Cancer Treat. Rep. *62:* 1937 (1978).

1298 Schultz, R. M.; Pavlidis, N. A.; Stylos, W. A.; Chirigos, M. A.: Regulation of macrophage tumoricidal function: A role for prostaglandins of the E series. Science *202:* 320 (1978).

1299 Schultze, H. E.; Schmidtberger, R.; Haupt, H.: Untersuchungen über die gebundenen Kohlenhydrate in isolierten Plasmaproteiden. Biochem. Z. *329:* 490 (1958).

1300 Schultze, H. E.; Heremans, J. F.: Molecular biology of human proteins, vol. 1, p. 246 (1966).

1301 Schwab, U.; Stein, H.; Gerdes, J.; Lemke, H.; Kirchner, H.; Schaadt, M.; Diehl, V.: Production of a monoclonal antibody specific for Hodgkin and Sternberg-Reed cells of Hodgkin's disease and a subset of normal lymphoid cells. Nature, Lond. *299:* 65 (1982).

1302 Schwartz, M.; Waltenbaugh, C.; Dorf, M.; Cesla, R.; Sela, M.; Benacerraf, B.: Determinants of antigenic molecules responsible for genetically controlled regulation of immune responses. Proc. natn. Acad. Sci. USA *73:* 2862 (1976).

1303 Schwartz, A.; Askenase, P. W.; Gershon, R. K.: Regulation of delayed-type hypersensitivity reactions by cyclophosphamide-sensitive T cells. J. Immun. *121:* 1573 (1978).

1304 Schwartz, B. D.; Cullen, S. E.: Chemical characteristics of Ia antigens. Semin. Immunopathol. *1:* 85 (1978).

1305 Schwartz, R. H.; Yano, A.; Paul, W. E.: Interaction between antigen presenting cells and primed T lymphocytes: An assessment of Ir gene expression in the antigen presenting cell. Immunol. Rev. *40:* 153 (1978).

1306 Schwarz, H.; Fischinger, P. J.; Ihle, J. N.; Thiel, H.-J.; Weiland, F.; Bolognesi, D. P.; Schäfer, W.: Properties of mouse leukemia virus. XVI. Suppression of spontaneous fatal leukemias in AKR mice by treatment with broadly reacting antibody against the viral glycoproteins gp 71. Virology *93:* 159 (1979).

1307 Schwarzenberg, L.; Mathé, G.; Schneider, M.; Amiel, J. L. A.; Schlumberger, J. R.: Attempted adoptive immunotherapy of acute leukaemia by leukocyte transfusions. Lancet *ii:* 365 (1966).

1308 Schweizer, K.; Gillissen, G.; Lutzeyer, W.: Immunologisch induzierte Hemmung und Förderung des Tumorwachstums (Ehrlich-Ascites-Tumor): Bedeutung verschiedener antigenhaltiger Präparationen. Med. Microbiol. Immunol. *159:* 251 (1974).

1309 Scott, M. T.: Corynebacterium parvum as a therapeutic anti-tumor agent in mice. II. Local injection. J. natn. Cancer Inst. *53:* 861 (1975).

1310 Scott, W. A.; Mahoney, E. M.; Cohn, Z. A.: Membrane lipids and the control of endocytosis; in van Furth, Mononuclear phagocytes, functional aspects, part II, p. 27 (Nijhoff, The Hague 1980).

1311 Scott, W. A.; Zrike, J. M.; Hamill, A. L.; Kempe, J.; Cohn, Z. A.: Regulation of arachidonic acid metabolites in macrophages. J. exp. Med. *152:* 324 (1980).

1312 Scott, W. A.; Pawlowski, N. A.; Murray, H. W.; Andreach, M.; Zrike, J.; Cohn, Z. A.: Regulation of arachidonic acid metabolism by macrophage activation. J. exp. Med. *155:* 1148 (1982).

1313 Sedlacek, H. H.: Pathophysiological aspects of immune complex diseases, part I. Klin. Wschr. *58:* 543 (1980).

1314 Sedlacek, H. H.: Pathophysiological aspects of immune complex diseases, part II. Klin. Wschr. *58:* 593 (1980).

1315 Sedlacek, H. H.: How to find immunomodulators – a look backward and forward. Behring Inst. Mitt. *74:* 122 (1984).

1316 Sedlacek, H. H.; Seiler, F. R.: Demonstration of vibrio cholerae neuraminidase (VCN) on the surface of VCN-treated cells. Behring Inst. Mitt. *55:* 254 (1974).

1317 Sedlacek, H. H.; Meesmann, H.; Seiler, F. R.: Regression of spontaneous mammary tumors in dogs after injection of neuraminidase-treated tumor cells. Int. J. Cancer *15:* 409 (1975).

1318 Sedlacek, H. H.; Dersjant, H.; Baudner, S.; Seiler, F. R.: Critical evaluation of the value of immunofluorescence methods for the demonstration of immunoglobulins on the lymphocyte surface. Behring Inst. Mitt. *59:* 38 (1976).

1319 Sedlacek, H. H.; Seiler, F. R.: The effect of vibrio cholerae neuraminidase (VCN) on the local and humoral immunological reaction of mice against SRBC. Z. Immun Forsch. *153:* 353 (1977).

1320 Sedlacek, H. H.; Seiler, F. R.; Schwick, H. G.: Neuraminidase and tumor immunotherapy. Klin. Wschr. *55:* 199 (1977).

1321 Sedlacek, H. H.; Johannsen, R.; Seiler, F. R.: Possible immunological action of vibrio cholerae neuraminidase (VCN) in tumor immunotherapy; in Griffith, Regamey, Develop. biol. Standard., vol. 38, p. 387 (Karger, Basel 1978).

1322 Sedlacek, H. H.; Seiler, F. R.: Effect of vibrio cholerae neuraminidase on the cellular immune response in vivo; in Rainer, Proc. Symp. Immunotherapy of malignant diseases, p. 268 (Schattauer, Stuttgart, New York 1978).

1323 Sedlacek, H. H.; Seiler, F. R.: Immunotherapy of neoplastic diseases with neuraminidase: Contradictions, new aspects, and revised concepts. Cancer Immunol. Immunother. *5:* 153 (1978).

1324 Sedlacek, H. H.; Seiler, F. R.: Neuraminidase immunotherapy of spontaneous canine mammary tumors; in Rainer, Proc. Symp. Immunotherapy of malignant diseases, p. 227 (Schattauer, Stuttgart, New York 1978).

1325 Sedlacek, H. H.; Seiler, F. R.: Spontaneous mammary tumors in mongrel dogs. A relevant model to demonstrate tumor therapeutical success by application of neuraminidase-treated tumor cells; in Griffith, Regamey, Develop. biol. Standard., vol. 38, p. 399 (Karger, Basel 1978).

1326 Sedlacek, H. H.; Lüben, G.; Bengelsdorff, H.-J.; Gronski, P.; Schmidt, K.-H.; Seiler, F. R.: Different biological distribution of immune complexes of varying antigen-antibody composition and lattice formation. Behring Inst. Mitt. *64:* 131 (1979).

1327 Sedlacek, H. H.; Weise, M.; Lemmer, A.; Seiler, F. R.: Immunotherapy of spontaneous mammary tumors in mongrel dogs with autologous tumor cells and neuraminidase. Cancer Immunol. Immunother. 6: 47 (1979).
1328 Sedlacek, H. H.; Bengelsdorff, H. J.; Seiler, F. R.: Minimal residual disease may be treated by chessboard vaccination with vibrio cholerae neuraminidase (VCN) and tumour cells; in Hellmann, Hilgaard, Eccles, Metastasis clinical and experimental aspects, p. 310 (Nijhoff, The Hague, Boston, London 1980).
1329 Sedlacek, H. H.; Hagmayer, G.; Seiler, F. R.: Tumor immunotherapy using the adjuvant neuraminidase: preliminary results of a randomized prospective study in the canine mammary tumor. In: Proc. Int. Symp. on Immunomodulation by Microbial Products and Related Synthetic Compounds, (Intl. Congress Series, No. 563) Osaka, July 1981.
1330 Sedlacek, H. H.; Seiler, F. R.: Problems in evaluating efficacy of immunopotentiators. In: Proc. Int. Symp. on Immunomodulation by Microbial Products and Related Synthetic Compounds (Intl. Congress Series, No. 563) Osaka, July 1981.
1331 Sedlacek, H. H.; Seiler, F. R.: Zur Frage der Pathogenität von Immunkomplexen und der intravenösen klinischen Anwendung von IgG und enzymatisch behandelten IgG-Präparationen; in Seiler, Geursen, Beitr. Infusionstherapie klin. Ernähr., vol. 9, p. 80 (Karger, Basel 1982).
1332 Sedlacek, H. H.; Weidmann, E.; Seiler, F. R.: Tumour immunotherapy using Vibrio Cholerae Neuraminidase (VCN); in Jelfaszewicz et al., Bacteria and cancer, p. 263 (Academic Press, London 1982).
1333 Sedlacek, H. H.; Schorlemmer, H. U.; Bosslet, K.; Dickneite, G.: Seiler, F. R.: Immunomodulation by Bestatin: Phenotypic description of its preclinical action and effectivity. Proc. 13th Int. Congress of Chemotherapy, Vienna (1983).
1334 Sedlacek, H. H.; Dickneite, G.; Schorlemmer, H. U.: Präklinische Prüfung von Immunmodulatoren unter spezieller Berücksichtigung von Chemoimmuntherapeutika. Akt. Onkol. 24: 52 (1985).
1335 Sedlacek, H. H.: Specific tumortherapy with neuraminidase and tumor cells. Mechanism, therapeutic potency and future outlook (1986).
1336 Sedlacek, H. H.; Hagmeyer, G.; Bengelsdorff, H. J.; Seiler, F. R.: Further studies on chessboard vaccinations for tumor therapy (in preparation 1986).
1337 Seeman, P.: Ultrastructure of membrane lesions in immune lysis, osmotic lysis and drug induced lysis. Fed. Proc. 33: 2116 (1974).
1338 Segall, A.; Weiler, O.; Genin, J.; Lacour, J.; Lacour, F.: In vitro study of cellular immunity against autochthonous human cancer. Int. J. Cancer 9: 417 (1972).
1339 Seigler, H. F.; Shingleton, W. W.; Metzgar, R. S.; Buckley, C. E.; Bergoc, P. M.; Miller, D. S.; Fetter, B. F.; Phaup, M. B.: Non-specific and specific immunotherapy in patients with melanoma. Surgery 72: 162: (1972).
1340 Seiler, F. R.; Sedlacek, H. H.: Alterations of immunological phenomena by neuraminidase: Marked rise in the number of lymphocytes forming rosettes or bearing immunoglobulin receptors. Behring Inst. Mitt. 55: 258 (1974).
1341 Seiler, F. R.; Sedlacek, H. H.; Lüben, G.; Wiegandt, H.: Alteration of the lymphocyte surface by Vibrio cholerae neuraminidase, gangliosides and lysolecithins. Behring Inst. Mitt. 59: 22 (1976).
1342 Seiler, F. R.; Sedlacek, H. H.: Chessboard vaccination: A pertinent approach to immunotherapy of cancer with neuraminidase and tumor cells? in Rainer, Proc.

of the Symp. Immunotherapy on Malignant Disease, Vienna, Nov. 9–12 (1977), p. 479 (Schattauer, Stuttgart 1978).
1343 Seiler, F. R.; Sedlacek, H. H.: BCG versus VCN: The antigenicity and the adjuvant effect of both compounds. Recent results in cancer research, vol. 75, p. 53 (Springer, Berlin, Heidelberg 1980).
1344 Seiler, F. R.; Gronski, P.; Kurrle, R.; Lüben, G.; Harthus, H.-P.; Ax, W.; Bosslet, K.; Schwick, H.-G.: Monoklonale Antikörper: Chemie, Funktion und Anwendungsmöglichkeiten. Angew. Chem. *3:* 141 (1985).
1345 Sendo, F.; Aoki, T.; Boyse, E. A.; Buofo, C. K.: Natural occurrence of lymphocytes showing cytotoxic activity to Balb/c radiation-induced leukemia RLol cells. J. natn. Cancer Inst. *55:* 603 (1975).
1346 Sener, S. F.; Brown, J. M.; Hyatt, C. L.: Serologic analysis of human tumor antigens. III. Reactivity of patients with melanoma and osteogenic sarcoma to cultured tumor cells and fibroblasts in the immune adherence assay. Cancer Immunol. Immunother. *11:* 243 (1981).
1347 Senik, A.; Gresser, I.; Maury, C.; Gidlund, M.; Orn, A.; Wigzell, H.: Enhancement by interferon of natural killer cell activity in mice. Cell Immunol. *44:* 186 (1979).
1348 Seon, B. K.; Negoro, S.; Minowada, J.; Yoshizaki, K.: Human T cell leukemia antigens on the cell membranes: purification, molecular characterization, and preparation of specific antisera. J. Immunol. *127:* 2580 (1981).
1349 Sercarz, E. E.; Yowell, R. L.; Turkin, D.; Miller, A.; Araneo, B. A.; Adorini, L.: Different functional specificity repertoires for suppressor and helper T cells. Immunol. Rev. *39:* 108 (1978).
1350 Serrou, B.; Cupissol, D.; Flad, H.: Phase I evaluation of Bestatin in patients bearing advanced solid tumors. Int. J. Immunopharmacol. *2:* 168 (1980).
1351 Seshadri, M.; Poduval, T. B.; Sundaram, K.: Studies on metastases. I. Role of sensitisation and immunosuppression. J. natn. Cancer Inst. *63:* 1205 (1979).
1352 Seshadri, M.; Poduval, T. B.: Immunity stimulation and metastasis. Cancer Immunol. Immunother. *9:* 213 (1980).
1353 Sethi, K. K.; Brandis, H.: Brief communication: Protection of mice from malignant tumor implants by enucleated tumor cells. J. natn. Cancer Inst. *53:* 1175 (1974).
1354 Seto, M.; Takahashi, T.; Nakamura, S.; Matsudaira, Y.; Nishizuka, Y.: In vivo antitumor effects of monoclonal antibodies with different immunoglobulin classes. Cancer Res. *43:* 4768 (1983).
1355 Sharma, B.: Generation of killer lymphocytes and adoptive immunotherapy for human cancer. Cancer Immunol. Immunother. *7:* 207 (1980).
1356 Shaw, S.; Pichler, W. J.; Nelson, D. L.: Fc receptors on human T-lymphocytes. III. Characterization of subpopulations involved in cell-mediated lympholysis and antibody-dependent cellular cytotoxicity. J. Immun. *122:* 599 (1979).
1357 Shearer, G. M.; Rehn, T. G.; Garbarino, C. A.: Cell mediated lympholysis of trinitro-phenyl-modified autologous lymphocytes effector cell specificity to modified cell surface components controlled by the H-2K and H-2D serological regions of the murine major histocompatibility complex. J. exp. Med. *141:* 1348 (1975).
1358 Shevach, E. M.; Paul, W. E.; Green, I.: Histocompatibility-linked immune response gene function in guinea pigs. Specific inhibition of antigen-induced lymphocyte proliferation by alloantisera. J. exp. Med. *136:* 1207 (1972).

1359 Shevach, E. M.; Rosenthal, A. S.: Function of macrophages in antigen recognition by guinea pig T lymphocytes. II. Role of the macrophage in the regulation of genetic control of the immune response. J. exp. Med. *138:* 1213 (1973).
1360 Shevach, E. M.; Green, I.; Paul, W. E.: Alloantiserum-induced inhibition of immune response gene product function. II. Genetic analysis of target antigens. J. exp. Med. *139:* 679 (1974).
1361 Shiku, H.; Takahashi, T.; Oettgen, H. F.; Old, L. J.: Cell surface antigens of human malignant melanoma. II. Serological typing with immune adherence assays and definition of two new antigens. J. exp. Med. *144:* 873 (1976).
1362 Shiku, H.; Takahashi, T.; Carey, T. E.; Resnick, L. A.; Oettgen, H. F.; Old, L. J.: Cell surface antigens of human cancer; in Ruddon, Leesburg, Biological markers of neoplasia: Basic and applied aspects, p. 73 (Elsevier, North Holland, Amsterdam 1978).
1363 Shin, H. S.; Hayden, M. L.; Gately, C. L.: Antibody-mediated suppression of lymphoma: Participation of platelets, lymphocytes, and nonphagocytic macrophages. Proc. natn. Acad. Sci. USA *71:* 163 (1974).
1364 Shin, H. S.; Johnson, R. J.; Pasternack, G. R.; Economou, J. S.: Mechanisms of tumor immunity: The role of antibody and nonimmune effectors; in Kallós, Waksman, de Weck, Prog. Allergy, vol. 25, p. 163 (1978).
1365 Shinitzky, M.: An efficient method for modulation of cholesterol level in cell membranes. FEBS Lett. *85:* 317 (1978).
1366 Shinitzky, M.; Barenholz, Y.: Dynamics of the hydrocarbon layer in liposomes of lecithin and sphingomyelin containing dicetylphosphate. J. biol. Chem. *249:* 2652 (1974).
1367 Shinitzky, M.; Inbar, M.: Difference in microviscosity induced by different cholesterol levels in the surface membrane lipid layer of normal lymphocytes and malignant lymphoma cells. J. molec. Biol. *85:* 603 (1974).
1368 Shinitzky, M.; Skornick, Y.; Haran-Ghera, N.: Effective tumor immunization induced by cells of elevated membrane lipid microviscosity. Proc. natn. Acad. Sci. USA *76:* 5313 (1979).
1369 Shirai, T. et al.: Cloning and expression in escherichia coli of the gene for human tumour necrosis factor. Nature *313:* 803 (1985).
1370 Sikora, K.; Phillips, J.: Human monoclonal antibodies to glioma cells. Br. J. Cancer *43:* 105 (1981).
1371 Sikora, K.; Wright, R.: Human monoclonal antibodies to lung-cancer antigens. Br. J. Cancer *43:* 696 (1981).
1372 Silberberg-Sinakin, I.; Thorbecke, G. J.; Baer, R. L.; Rosenthal, S. A.; Berezowsky, V.: Antigen-bearing Langerhans cells in the skin dermal lymphatics and lymph nodes. Cell. Immunol. *25:* 137 (1976).
1373 Silva, A.; Bonavida, B.; Targan, S.: Mode of action of interferon-mediated modulation of natural killer cytotoxic activity: recruitment of pre-NK cells and enhanced kinetics of lysis. J. Immun. *125:* 479 (1980).
1374 Silverstein, S. C.; Steinman, R. M.; Cohn, Z. A.: Endocytosis. Annu. Rev. Biochem. *46:* 669 (1977).
1375 Simes, R. J.; Kearney, R.; Nelson, D. S.: Role of a non-committed accessory cell in the in vivo suppression of a syngeneic tumour by immune lymphocytes. Immunology *29:* 343 (1975).

1376 Simmons, R. L.; Lipschultz, M. L.; Rios, A.; Ray, P. K.: Failure of neuraminidase to unmask histocompatibility antigens on trophoblast. Nature new Biol. *231:* 111 (1971).
1377 Simon, H. B.; Sheagren, J. N.: Enhancement of macrophage bactericidal capacity by antigenically stimulated immune lymphocytes. Cell. Immunol. *4:* 163 (1972).
1378 Singer, A.; Dickler, H. B.; Hodes, R. J.: Cellular and genetic control of antibody responses in vitro. II. IR gene control of primary IgM responses to trinitrophenyl-conjugates of poly-L-(tyr, glu)-poly-D, L-ala-poly-L-lys and poly-L-(his, glu)-poly-D, L-ala-poly-L-lys. J. exp. Med. *146:* 1096 (1977).
1379 Singer, A.; Cowing, C.; Hathcock, K. S.; Dickler, H. B.; Hodes, R. J.: Cellular and genetic control of antibody responses in vitro. III. Immune response gene regulation of accessory cell function. J. exp. Med. *147:* 1611 (1978).
1380 Sinkovics, J. G.: Immunotherapy with viral oncolysates for sarcoma. JAMA *237:* 869 (1977).
1381 Sinkovics, J. G.: Bacterial products in the immunotherapy of human tumours; in Jeljaszewicz et al., Bacteria and cancer, p. 359 (Academic Press, London, New York 1982).
1382 Sjögren, H. O.: Studies on specific transplantation resistance to polyoma-virus-induced tumors. I. Transplantation resistance induced by polyoma virus infection. J. natn. Cancer Inst. *32:* 361 (1964).
1383 Sjögren, H. O.: Studies on specific transplantation resistance to polyoma-virus-induced tumors. II. Mechanism of resistance induced by polyoma virus infection. J. natn. Cancer Inst. *32:* 375 (1964).
1384 Sjögren, H. O.: Studies on specific transplantation resistance to polyoma-virus-induced tumors. III. Transplantation resistance to genetically compatible polyoma tumors induced by polyoma tumor homografts. J. natn. Cancer Inst. *32:* 645 (1964).
1385 Sjögren, H. O.: Transplantation methods as a tool for detection of tumor-specific antigens. Prog. exp. Tumor Res., vol. 6, p. 289 (Karger, Basel 1965).
1386 Sjögren, H. O.; Hellström, I.; Klein, G.: Transplantation of polyoma virus-induced tumors in mice. Cancer Res. *21:* 329 (1961).
1387 Sjögren, H. O.; Jonsson, N.: Resistance against isotransplantation of mouse tumors induced by rous sarcoma virus. Expl Cell Res. *32:* 618 (1963).
1388 Sjögren, H. O.; Hellström, I.; Bansal, S. C.: Suggestive evidence that the «Blocking Antibodies» of tumor-bearing individuals may be antigen-antibody complexes. Proc. natn. Acad. Sci. USA *68:* 1372 (1971).
1389 Slanetz, C. A., Jr.; McCollester, D. L.; Kanor, S.: Autologous anticancer antigen preparation for specific immunotherapy in advanced cancer patients: A phase I clinical trial. Cancer Immunol. Immunother. *13:* 75 (1982).
1390 Small, M.; Trainin, N.: Separation of populations of sensitized lymphoid cells into fractions inhibiting and fractions enhancing syngeneic tumor growth in vivo. J. Immunol. *117:* 292 (1976).
1391 Smets, L. A.; Van Beek, W. P.: Carbohydrates of the tumor cell surface. Biochim. biophys. Acta *738:* 237 (1984).
1392 Smith, T. J.; Wagner, R. R.: Rabbit macrophage interferons. I Conditions for biosynthesis by virus-infected and uninfected cells. J. exp. Med. *125:* 559 (1967)
1393 Smith, D. F.; Walborg, E. F.: Isolation and chemical characterization of cell surface

sialoglycopeptide fractions during progression of rat ascites hepatoma AS-30D. Cancer Res. *32:* 543 (1972).

1394 Smith, H. G.; Meltzer, M. S.; Leonard, E. J.: Suppression of tumor growth in strain 2 guinea pigs by xenogeneic antitumor antibody. J. natn. Cancer Inst. *57:* 809 (1976).

1395 Smith, K. A.; Gilbride, K. J.; Favata, M. F.: Interleukin 1-promoted interleukin 2 production. Behring Inst. Mitt. *67:* 4 (1980).

1396 Smyth, H.; Farrell, D. J.; O'Kennedy, R.; Corrigan, A.: Transplantability of neuraminidase-treated ascites tumour cells. Eur. J. Cancer *13:* 1313 (1977).

1397 Snell, G. D.; Cloudman, A. M.; Failor, E.; Douglass, P.: Inhibition and stimulation of tumor homoiotransplants by prior injections of lyophilized tumor tissue. J. natn. Cancer Inst. *6:* 303 (1946).

1398 Snell, G. D.; Cloudman, A. M.; Woodworth, E.: Tumor immunity in mice, induced with lyophilized tissue, as influenced by tumor strain, host strain, source of tissue, and dosage. Cancer Res. *8:* 429 (1948).

1399 Snodgrass, M. J.; Hanna, M. G.: Ultrastructural studies of histiocyte-tumor cell interactions during tumor rejection mediated by intralesional injection of Mycobacterium bovis. Cancer Res. *33:* 701 (1973).

1400 Snyder, H. W., Jr.; Hardy, W. D., Jr.; Zuckerman, E. E.; Fleissner, E.: Characterisation of a tumour-specific antigen on the surface of feline lymphosarcoma cells. Nature, Lond. *275:* 656 (1978).

1401 Snyderman, R.; Pike, M. C.; Altman, L. C.: Abnormalities of leukocyte chemotaxis in human disease. Ann. N. Y. Acad. Sci. *256:* 386 (1975).

1402 Snyderman, R.; Pike, M. C.: An inhibitor of macrophage chemotaxis produced by neoplasms. Science *192:* 370 (1976).

1403 Snyderman, R.; Pike, M. C.; Blaylock, B. L.; Weinstein, P.: Effects of neoplasms on inflammation: Depression of macrophage accumulation after tumour implantation. J. Immun. *116:* 585 (1976).

1404 Snyderman, R.; Meadows, L.; Holder, W.; Wells, S., Jr.: Abnormal monocyte chemotaxis in patients with breast cancer: Evidence for a tumour-mediated effect. J. natn. Cancer Inst. *60:* 737 (1978).

1405 Sobol, R. E.; Dillman, R. O.; Smith, J. D.: Phase I evaluation of murine monoclonal anti-melanoma antibody in man: Preliminary observations; in Mitchell, Oettgen, Hybridoma and cancer diagnosis and treatment, p. 199 (Raven Press, New York 1982).

1406 Solbach, W.; Röllinghoff, M.; Wagner, H.: Die Rolle von Interleukin-2 bei der Aktivierung von zytotoxischen T-Lymphozyten. Klin. Wschr. *61:* 67 (1983).

1407 Sone, S.; Poste, G.; Fidler, I. J.: Rat alveolar macrophages are susceptible to activation by free and liposome-encapsulated lymphokines. J. Immunol. *124:* 2197 (1980).

1408 Song, C. W.; Levitt, S. H.: Immunotherapy with neuraminidase treated tumour cells after radiotherapy. Radiat. Res. *64:* 485 (1975).

1409 Sorg, C.; Neumann, C.; Klimetzek, V.; Hannich, D.: Lymphokine-induced modulation of macrophge functions; in van Furth, Mononuclear phagocytes, functional aspects, p. 520 (Nijhoff, The Hague 1980).

1410 Souter, R. G.; Gill, P. G.; Gunning, A. J.; Morris, P. J.: Failure of specific active immunotherapy in lung cancer. Br. J. Cancer *44:* 496 (1981).

1411 Sparks, F.C.; Breeding, J.H.: Tumor regression and enhancement resulting from immunotherapy with bacillus Calmette-Guérin and neuraminidase. Cancer Res. *34:* 3262 (1974).
1412 Spellman, C.W.; Daynes, R.A.: Modification of immunological potential by ultraviolet radiation. II. Generation of suppressor cells in short-term UV-irradiated mice. Transplantation *24:* 120 (1977).
1413 Spiegel, R.J.: Magrath, I.T.; Shutta, J.A.: Role of cytoplasmic lipids in altering diphenylhexatriene fluorescence polarization in malignant cells. Cancer Res. *41:* 452 (1981).
1414 Spitalny, G.L.; North, R.J.: Subversion of host defense mechanisms by malignant tumours: An established tumour as a privileged site for bacterial growth. J. exp. Med. *145:* 1264 (1977).
1415 Spreafico, F.; Vecchi, A.; Mantovani, A.; Poggi, A.; Franchi, G.; Anaclerio, A.; Garrattini, S.: Characterisation of the immunostimulants – Levamisole and Tetramisole. Eur. J. Cancer *11:* 555 (1975).
1416 Spreafico, F.; Mantovani, A.: Immunomodulation by cancer chemotherapeutic agents and antineoplastic activity. Pathobiol. Annu. *11:* 177 (1981).
1417 Sprent, J.: Role of H-2 gene products in the function of T helper cells from normal and chimeric mice in vivo. Immunol. Rev. *42:* 108 (1978).
1418 Springer, G.F.: Blood-group and Forssman antigenic determinants shared between microbes and mammalian cells. Prog. Allergy, vol. 15, p. 9 (Karger, Basel 1971).
1419 Springer, G.F.; Desai, P.R.; Banatwala, I.: Blood group MN-specific substances and precursors in normal and malignant human breast tissues. Naturwissenschaften *61:* 457 (1974).
1420 Springer, G.F.; Desai, P.R.: Depression of Thomsen-Friedenreich (anti-T) antibody in humans with breast carcinoma. Naturwissenschaften *62:* 302 (1975).
1421 Springer, G.F.; Desai, P.R.: Human blood group MN and precursor specificities: Structural and biological aspects. Carbohyd. Res. *40:* 183 (1975).
1422 Springer, G.F.; Desai, P.R.: Increase in anti-T titer scores of breast-carcinoma patients following mastectomy. Naturwissenschaften *62:* 587 (1975).
1423 Springer, G.F.; Desai, P.R.; Banatwala, I.: Brief communication: Blood group MN antigens and precursors in normal and malignant human breast glandular tissue. J. natn. Cancer Inst. *54:* 335 (1975).
1424 Springer, G.F.; Desai, P.R.; Yang, H.J.; Schachter, H.; Narasimhan, S.: Interrelations of blood group M and precursor specificities and their significance in human carcinoma; in Mohn, Plunkett, Cunningham, Lambert, Human blood groups, p. 179 (Karger, Basel 1977).
1425 Springer, G.F.; Desai, P.R.; Murthy, M.S.; Scanlon, E.F.: Delayed-type skin hypersensitivity reaction (DTH) to Thomsen-Friedenreich (T) antigen as diagnostic test for human breast adenocarcinoma. Klin. Wschr. *57:* 961 (1979).
1426 Springer, G.F.; Desai, P.R.; Murthy, M.S.; Tegtmeyer, H.; Scanlon, E.F.: Human carcinoma-associated precursor antigens of the blood group MN system and the host's immune responses to them. Prog. Allergy, vol. 26, p. 42 (Karger, Basel 1979).
1427 Springer, G.F.; Desai, P.R.; Murthy, M.S.; Yang, H.J.; Scanlon, E.F.: Precursors of the blood group MN antigens as human carcinoma-associated antigens. Transfusion *19:* 233 (1979).
1428 Springer, G.F.; Desai, P.R.; Fry, W.A.; Goodale, R.L.; Shearen, J.G.; Scanlon,

E. F.: Importance of breast-, lung-, and pancreas carcinoma cell membrane T antigen in their immunodiagnosis; in Galeotti et al., Membranes in tumour growth, p. 575 (Elsevier, North Holland 1982).

1429 Springer, G. F.; Murthy, S. M.; Desai, P. R.; Fry, W. A.; Tegtmeyer, H.; Scanlon, E. F.: Patients' immune response to breast and lung carcinoma-associated Thomsen-Friedenreich (T) specificity. Klin. Wschr. *60:* 121 (1982).

1430 Springer, G. F.; Desai, P. R.; Fry, W. A.; Goodale, R. L.; Shearen, J. G.; Scanlon, E. F.: T antigen, a tumor marker against which breast, lung, pancreas carcinoma patients mount immune responses. Cancer Detect. Prevent. *6:* 111 (1983).

1431 Springer, G. F.; Taylor, C. R.; Howard, D. R.; Tegtmeyer, H.; Desai, P. R.; Murthy, S. M.; Felder, B.; Scanlon, E. F.: Tn, a carcinoma-associated antigen, reacts with anti-Tn of normal human sera. Cancer *55:* 561 (1985).

1432 Sreevalsan, T.; Taylor-Papadimitriou, J.; Rozengurt, E.: Selective inhibition by interferon of serum-stimulated biochemical events in 3T3 cells. Biochem. biophys. Res. Commun. *87:* 679 (1979).

1433 Staab, H.-J.; Anderer, F. A.: Immunogenicity of tumour cells modified with various chemicals. Br. J. Cancer *35:* 395 (1977).

1434 Staab, H.-J.; Anderer, F. A.: Chemical modification and immunogenicity of membrane fractions from mouse tumour cells. Br. J. Cancer *38:* 496 (1978).

1435 Stackpole, C. W.: Modulation of thymus-leukemia antigens on mouse leukemia cells induced by IgG, but not IgM, antibody. J. natn. Cancer Inst. *64:* 917 (1980).

1436 Stadecker, M. J.; Calderch, J.; Karnovsky, M. L.; Unanue, E. R.: Synthesis and release of thymidine by macrophages. J. Immun. *119:* 1738 (1977).

1437 Starling, J. J.; Sieg, S. M.; Beckett, M. L.; Schellhammer, P. F.; Ladaga, L. E.; Wright, G. L., Jr.: Monoclonal antibodies to human prostate and bladder tumor-associated antigens. Cancer Res. *42:* 3084 (1982).

1438 Steele, G., Jr.; Sjögren, H. O.: Embryonic antigens associated with chemically induced colon carcinomas in rats. Int. J. Cancer *14:* 435 (1974).

1439 Steele, R. W.; Myers, M. G.; Vincent, M. M.: Transfer factor for the prevention of varicella-zoster infection in childhood leukemia. New Engl. J. Med. *303:* 355 (1980).

1440 Steinman, R. M.; Cohn, Z. A.: A novel adherent cell in mouse lymphoid organs; in Rosenthal, Immune recognition, p. 571 (Academic Press, New York 1975).

1441 Steinman, R. M.; Witmer, M. D.: Lymphoid dendritic cells are potent stimulators of the primary mixed leukocyte reaction in mice. Proc. natn. Acad. Sci. USA *75:* 5132 (1978).

1442 Steinman, R. M.; Nussenzweig, M. C.: Dendritic cells: Features and functions. Immunol. Rev. *53:* 127 (1980).

1443 Stephenson, J. R.; Essex, M.; Hino, S.; Hardy, W. D., Jr.; Aaronson, S. A.: Feline oncornavirus-associated cell-membrane antigen (FOCMA): distinction between FOCMA and the major virion glycoprotein. Proc. natn. Acad. Sci USA *74:* 1219 (1977).

1444 Steplewski, Z.; Chang, T. H.; Herlyn, M.; Koprowski, H.: Release of monoclonal antibody-defined antigens by human colorectal carcinoma and melanoma cells. Cancer Res. *41:* 2723 (1981).

1445 Steplewski, Z.; Lubeck, M. D.; Koprowski, H.: Human macrophages armed with murine immunoglobulin G2a antibodies to tumors destroy human cancer cells. Science *221:* 865 (1983).

1446 Stern, P. L.; Willison, K. R.; Lennox, E.; Galfre, G.; Milstein, C.; Secher, D. S.; Ziegler, A.; Springer, T.: Monoclonal antibodies as probes for differentiation and tumor-associated antigens. A Forssman specificity on teratocarcinoma stem cells. Cell *14:* 775 (1978).

1447 Stern, P.; Gidlund, M.; Örn, A.; Wigzell, H.: Natural killer cells mediate lysis of embryonal carcinoma cells lacking MHC. Nature *285:* 341 (1980).

1448 Stewart, W. E., II.; Gresser, I.; Tovey, M. G.; Bandu, M. T.; Le Gof, S.: Identification of the cell multiplication inhibitory factors in interferon preparations as interferons. Nature, Lond. *262:* 300 (1976).

1449 Stewart, T. H. M.; Hollinshead, A. C.; Harris, J. E.; Raman, S.: Specific active immunotherapy in lung cancer: The induction of long-lasting cellular responses to tumor-associated antigens. Recent Results Cancer Res. *80:* 232 (1982).

1450 Stingl, G.; Katz, S. I.; Clement, L.; Green, I.; Shevach, E. M.: Immunologic functions of Ia-bearing epidermal Langerhans cells. J. Immun. *121:* 2005 (1978).

1451 Stjernswärd, J.; Vanky, F.: Stimulation of lymphocytes by autochthonous cancer. Natn. Cancer Inst. Monogr. *35:* 237 (1972).

1452 Stuhlmiller, G. M.; Seigler, H. F.: Characterization of a chimpanzee anti-human melanoma antiserum. Cancer Res. *35:* 2132 (1975).

1453 Stutman, O.: Immunodeficiency and cancer; in Green, Cohen, Mechanisms of tumor immunity, p. 27 (John Wiley & Sons, New York 1977).

1454 Stutman, O.; Dien, P.; Wisum, R. E.; Lattime, E. C.: Natural cytotoxic cells against solid tumors in mice: blocking of cytotoxicity by D-mar nose. Proc. natn. Acad. Sci. USA *77:* 2895 (1980).

1455 Stutman, O.; Lattime, E.; Dien, P.; Cuttito, M.; Wisun, R.: Natural cytotoxic (NC) cells against solid tumors in mice: Inhibition of cytotoxicity by some simple sugars (Abstract 4654) Fed. Proc. *39:* 1151 (1980).

1456 Stutman, O.; Lattime, E. C.; Gillis, S.; Miller, R. A.: Mode of action of murine interleukin 2: Role of interleukin 2 in postthymic maturation of T cells. Behring Inst. Mitt. *67:* 95 (1980).

1457 Sugarbaker, E. V.; Egan, J. E.: Alterations of tumor specific transplantation antigens in metastases. Proc. Am. Ass. Cancer Res. *296:* 74 (1971).

1458 Sugiura, K.; Stock, C. C.: Studies in a tumor spectrum III. The effect of phosphoramides on the growth of a variety of mouse and rat tumours. Cancer Res. *15:* 38 (1955).

1459 Sumner, W. C.; Foraker, A. G.: Spontaneous regression of human melanoma. Clinical and experimental studies. Cancer, N. Y. *13:* 179 (1960).

1460 Sundsmo, J. S.; Müller-Eberhard, H. J.: Neoantigen of the complement membrane attack complex on cytotoxic human peripheral blood lymphocytes. J. Immunol. *122:* 2371 (1979).

1461 Sunshine, G. H.; Katz, D. R.; Feldman, M.: Dendritic cells induce proliferation to synthetic antigens under Ir gene control. J. exp. Med. *152:* 1817 (1980).

1462 Suter, L.; Bröcker, E.-B.; Brüggen, J.; Ruiter, D. J.; Sorg, C.: Heterogeneity of primary and metastatic human malignant melanoma as detected with monoclonal antibodies in cryostat sections of biopsies. Cancer Immunol. Immunother. *16:* 53 (1983).

1463 Svedersky, L. P.: Immunoregulatory lymphokines. I. Human lymphotoxin induction of interferon activity. Int. J. Immunopharmacol. *4:* 374 (1982).

1464 Svennevig, J.-L.; Lövik, M.; Svaar, H.: Isolation and characterization of lymphocytes and macrophages from solid, malignant human tumours. Int. J. Cancer *23:* 626 (1979).
1465 Svennevig, J.-L.; Svaar, H.: Content and distribution of macrophages and lymphocytes in solid malignant human tumours. Int. J. Cancer *24:* 754 (1979).
1466 Symes, M. O.; Riddell, A. G.; Immelman, E. J.; Terblanche, J.: Immunologically competent cells in the treatment of malignant disease. Lancet *i:* 1054 (1968).
1467 Szendröi, Z.; Balogh, F.: Der Prostata-Krebs (Akadémiai Kiadó, Budapest, 1965).
1468 Tachibana, T.; Klein, E.: Detection of cell surface antigens on monolayer cells. I. The application of immune adherence on a microscale. Immunology *19:* 771 (1970).
1469 Tada, T.; Okumura, K.: The role of antigen-specific T cell factors in the immune response. Adv. Immunol. *28:* 1 (1979).
1470 Tadakuma, T.; Pierce, C. W.: Mode of action of a soluble immune response suppressor (SIRS) produced by concanavalin A-activated spleen cells. J. Immun. *120:* 481 (1978).
1471 Takasugi, M.; Klein, E.: A microassay for cell-mediated immunity. Transplantation *9:* 219 (1970).
1472 Takasugi, M.; Mickey, M. R.; Terasaki, P. I.: Reactivity of lymphocytes from normal persons on cultured tumor cells. Cancer Res. *33:* 2898 (1973).
1473 Takasugi, M.; Akira, D.; Kinoshita, K.: Granulocytes as effectors in cell-mediated cytotoxicity of adherent target cells. Cancer Res. *35:* 2169 (1975).
1474 Takasugi, M.; Ramseyer, A.; Takasugi, J.: Decline of natural nonselective cell-mediated cytotoxicity in patients with tumor progression. Cancer Res. *37:* 413 (1977).
1475 Tallberg, T.: Cancer-immunotherapy by means of polymerised autologous tumour tissue with special reference to some patients with pulmonary tumour. Scand. J. resp. Dis., suppl. 89, p. 107 (1974).
1476 Tallberg, T.; Tykkä, H.; Halttunen, P.; Mahlberg, K.; Uusitalo, R.; Uusitalo, H.; Carlson, O.; Sandstedt, B.; Oravisto, K. J.; Lehtonen, T.; Sarna, S.; Strandström, H.: Cancer immunity. The effect in cancer – immunotherapy of polymerised autologous tumour tissue and supportive measures. Scand. J. clin. Lab. Invest. *39:* suppl. 151, p. 306 (1979).
1477 Talmadge, J. E.; Meyers, K. M.; Prieur, D. J.; Starkey, J. R.: Role of NK cells in tumor growth and metastasis in beige mice. Nature, Lond. *284:* 622 (1980).
1478 Talmadge, J. E.: Fidler, I. J.; Oldham, R. K.: The NCI preclinical screen of biological response modifiers. Behring Inst. Mitt. *74:* 189 (1984).
1479 Talmadge, E.; Lenz, B. F.; Collins, M. S.; Uithoven, K. A.; Schneider, M. A.; Adams, J. S.; Pearson, J. W.; Agee, W. J.; Fox, R. E.; Oldham, R. K.: Tumor models to investigate the therapeutic efficiency of immunomodulators. Behring Inst. Mitt. *74:* 219 (1984).
1480 Tamerius, J. D.; Garrigues, H. J.; Hellström, K. E.; Hellström, I.: An isotope-release assay and a terminal-labeling assay for measuring cell-mediated allograft and tumor immunity to small numbers of adherent target cells. J. immunol. Methods *22:* 1 (1978).
1481 Taniyama, T.; Holden, H. T.: Cytolytic activity of macrophages isolated from primary murine sarcoma virus (MSV) – induced tumours. Int. J. Cancer *24:* 151 (1979).

1482 Tauber, A. I.; Babior, B. M.: Evidence for hydroxyl radical production by human neutrophils. J. clin. Invest. *60:* 374 (1977).
1483 Taverne, J.; Matthews, N.; Depledge, P.; Playfair, J. H.L.: Malarial parasites and tumour cells are killed by the same component of tumour necrosis serum. Clin. exp. Immunol. *57:* 293 (1984).
1484 Teramoto, Y. A.; Mariani, R.; Wunderlich, D.; Schlom, J.: The immunohistochemical reactivity of a human monoclonal antibody with tissue sections of human mammary tumors. Cancer *50:* 241 (1982).
1485 Termijtelen, A.; Van Leeuwen, A.; Van Rood, J. J.: HLA-linked lymphocyte activating determinants. Immunol. Rev. *66:* 79 (1982).
1486 Terry, W. D.; Hodes, R. J.; Rosenberg, S. A.; Fisher, R. I.; Makuch, R.; Gordon, H. G.; Fisher, S. G.: Treatment of stage I and II malignant melanoma with adjuvant immunotherapy of chemotherapy: Preliminary analysis of a prospective, randomized trial; in Terry, Windhorst, 2nd Int. Conf. Immunotherapy of Cancer: Present status of trials in man, p. 47 (Raven Press, New York 1980).
1487 Terry, W. D.; Hodes, R. J.: Newer methods of cancer treatment. Section 1: Immunotherapy; in de Vita et al., Cancer principles and practice of oncology, chapter 48, p. 1788 (Lippincott Company, Philadelphia, Toronto 1982).
1488 Tevethia, S. S.; McMillan, V. L.; Kaplan, P. M.: Variation in immunosensitivity of SV40-transformed hamster cells. J. Immun. *106:* 1295 (1971).
1489 Tew, J. G.; Phipps, R. P.; Mandfel, T. E.: The maintenance and regulation of the humoral immune response: Persisting antigen and the role of follicular antigen-binding dendritic cells as accessory cells. Immunol. Rev. *53:* 175 (1980).
1490 Thakrai, K. K.; Goodson, W. H.; Hunt, T. K.: Stimulation of wound blood vessel, growth by wound macrophages. J. surg. Res. *26:* 430 (1979).
1491 Thatcher, N.; Hashmi, K.; Chang, J.; Swindell, R.; Crowther, D.: Anti-T antibody in malignant melanoma patients: Influence of response and survival following chemotherapy – changes in serum levels following C. parvum, BCG immunization. Cancer: *46:* 1378 (1980).
1492 Thiel, E.: Monoclonal antibodies against differentiation antigens of lymphopoiesis. Blut *47:* 247 (1983).
1493 Thierfelder, S.: Hemopoietic stem cells of rats but not of mice express Thy-1.1 alloantigen. Nature, Lond. *269:* 691 (1977).
1494 Thomas, L.: Discussion; in Von Lawrence, Cellular and humoral aspects of the hypersensitive state, p. 529 (Hoeber, New York 1959).
1495 Thomas, D. B.; Winzler, R. J.: Structural studies on human erythrocyte glycoproteins. Alkalilabile oligosaccharides. J. biol. Chem. *244:* 5943 (1969).
1496 Thomsen, O.: Ein vermehrungsfähiges Agens als Veränderer des isoagglutinatorischen Verhaltens der roten Blutkörperchen, eine bisher unbekannte Quelle der Fehlbestimmungen. Z. ImmunForsch. *52:* 85 (1927).
1497 Thomson, D. M. P.; Gold, P.; Freedman, S. O.: The isolation and characterization of tumor-specific antigens of rodent and human tumors. Cancer Res. *36:* 3518 (1976).
1498 Thomson, D. M. P.; Rauch, J. E.; Weatherhead, J. C.; Frielander, P.; O'Connor, R.; Grosser, N.; Shuster, J.; Gold, P.: Isolation of human tumour-specific antigens associated with β2-microglobulin. Br. J. Cancer *37:* 753 (1978).
1499 Thomson, A. W.; Cruickshank, N.; Fowler, E. F.: Fc receptor-bearing and phago-

cytic cells in syngeneic tumours of C. parvum and carrageenan-treated mice. Br. J. Cancer 39: 598 (1979).
1500 Thorsby, E.; Berle, E.; Nousiainen, H.: HLA-D region molecules restrict proliferative T cell responses to antigen. Immunol. Rev. 66: 39 (1982).
1501 Tidman, N.; Janossy, G.; Bodger, M.; Granger, S.; Kung, P. C.; Goldstein, G.: Delineation of human thymocyte differentiation pathways utilizing double-staining techniques with monoclonal antibodies. Clin. exp. Immunol. 45: 457 (1981).
1502 Timonen, T.; Ortaldo, J. R.; Herberman, R. B.: Characteristics of human large granular lymphocytes and relationship to natural killer and K cells. J. exp. Med. 155: 569 (1981).
1503 Ting, C. C.; Herberman, R. B.: Detection of tumor specific cell surface antigen of SV-40 induced tumors by the isotopic antiglobulin technique. Int. J. Cancer 7: 499 (1971).
1504 Ting, C. C.; Herberman, R. B.: Inverse relationship of polyoma tumor specific cell surface antigen to H-2 histocompatibility antigens. Nature new Biol. 232: 118 (1971).
1505 Ting, C. C.; Lavrin, D. H.; Shiu, G.; Herberman, R. B.: Expression of fetal antigens in tumor cells. Proc. natn. Acad. Sci. USA 69: 1664 (1972).
1506 Ting, C. C.; Ortaldo, J. R.; Herberman, R. B.: Expression of fetal antigens and tumor specific antigens in SV40 transformed cells. I. Serological analysis of the antigenic specificities. Int. J. Cancer 12: 511 (1973).
1507 Ting, C. C.; Rodrigues, D.; Herberman, R. B.: Expression of fetal antigens and tumor specific antigens in SV40 transformed cells. II. Tumor transplantation studies. Int. J. Cancer 12: 519 (1973).
1508 Ting, C. C.; Law, L. W.: Studies of H-2 restriction in cell-mediated cytotoxicity and transplantation immunity to leukemia-associated antigens. J. Immun. 118: 4 (1977).
1509 Ting, C. C.; Rodrigues, D.; Ting, R. C.; Wivel, N.; Collins, M. J.: Suppression of T cell-mediated immunity by tumor cells : Immunogenicity versus immunosuppression and preliminary characterization of the suppressive factors. Int. J. Cancer 24: 644 (1979).
1510 Tomecki, J.: The influence of immunization of Syrian hamsters with tumor cells treated with glutaraldehyde on transplantation immunity and the cytotoxic effect of lymphocytes on polyoma tumor cells. Arch. Immunol. Therap. Exp. 27: 209 (1979).
1511 Toy, J. L.: The interferons. Clin. exp. Immunol. 54: 1 (1983).
1512 Tracey, D. E.: The requirement for macrophages in the augmentation of natural killer cell activity by BCG. J. Immun. 123: 840 (1979).
1513 Tracey, D. E.; Adkinson, N. F., Jr.: Prostaglandin synthesis inhibitors potentiate the BCG-induced augmentation of natural killer cell activity. J. Immun. 125: 136 (1980).
1514 Treves, A. J.: In vitro induction of cell-mediated immunity against tumor cells by antigen-fed macrophages. Immunol. Rev. 40: 205 (1978).
1515 Treves, A. J.; Carnaud, C.; Trainin, N.: Enhancing T lymphocytes from tumour-bearing mice suppress host resistance to a syngeneic tumour. Eur. J. Immunol. 4: 722 (1974).
1516 Treves, A. J.; Cohen, I. R.; Feldman, M.: A syngeneic metastatic tumor model in

mice: The natural immune response of the host and its manipulation. Israel J. med. Scis *12:* 4 (1976).
1517 Treves, A. J.; Schechter, B.; Cohen, I. R.; Feldman, M.: Sensitization of T lymphocytes in vitro by syngeneic macrophages fed with tumor antigens. J. Immun. *116:* 1059 (1976).
1518 Trinchieri, G.; Aden, D. P.; Knowles, B. B.: Cell mediated cytotoxicity to SV40-specific tumour-associated antigens. Nature, Lond. *261:* 312 (1976).
1519 Trinchieri, G.; Santoli, D.: Antiviral activity induced by culturing lymphocytes with tumor-derived or virus-transformed cells. Enhancement of human natural killer cell activity by interferon and antagonistic inhibition of susceptibility of target cells to lysis. J. exp. Med. *147:* 1314 (1978).
1520 Trinchieri, G.; Santoli, D.; Dee, R. R.; Knowles, B. B.: Antiviral activity induced by culturing lymphocytes with tumor-derived or virus-transformed cells. Identification of the anti-viral activity as interferon and characterization of the human effector lymphocyte subpopulation. J. exp. Med. *147:* 1299 (1978).
1521 Trinchieri, G.; Santoli, D.; Koprowski, H.: Spontaneous cell-mediated cytotoxicity in humans: Role of interferon and immunoglobulins. J. Immun. *120:* 1849 (1978).
1522 Trinchieri, G.; Perussia, B.; Santoli, D.: Spontaneous cell-mediated cytotoxicity: Modulation by interferon; in Herberman, Natural cell-mediated immunity against tumors, p. 655 (Academic Press, New York 1980).
1523 Trinchieri, G.; Granato, D.; Perussia, B.: Interferon-induced resistance of fibroblasts to cytolysis mediated by natural killer cells: specificity and mechanism of the phenomenon. J. Immun. *126:* 335 (1981).
1524 Trinchieri, G.; Santoli, D.; Granato, D.; Perussia, B.: Antagonistic effects of interferons on the cytotoxicity mediated by natural killer cells. Fed. Proc. *40:* 2705 (1981).
1525 Trojanowski, J. Q.; Lee, V.; Pillsbury, N.; Lee, S.: Neuronal origin of human esthesioneuroblastoma demonstrated with anti-neurofilament monoclonal antibodies. New Engl. J. Med. *307:* 159 (1982).
1526 Trowbridge, I. S.; Domingo, D. L.: Antitransferrin receptor monoclonal antibody and toxin-antibody conjugates affect growth of human tumor cells. Nature, Lond. *294:* 171 (1981).
1527 Tsan, M. F.; McIntyre, P. A.: The requirement for membrane sialic acid in the stimulation of superoxide production during phagocytosis by human polymorphonuclear leukocytes. J. exp. Med. *143:* 1308 (1976).
1528 Tsan, M. F.; Douglass, K. H.; McIntyre, P. A.: Hydrogen peroxide production and killing of Staphylococcus aureus by human polymorphonuclear leukocytes. Blood *49:* 437 (1977).
1529 Tsuchida, T.; Fujiwara, H.; Tsuji, Y.; Hamaoka, T.: Inhibitory effects of propionibacterium-acnes-activated macrophages on the tumor metastasis enhanced by tumor-specific immunosuppression. Gann *72:* 205 (1981).
1530 Tuttle, R. L.; North, R. J.: Mechanisms of antitumor action of Corynebacterium parvum: nonspecific tumor cell destruction at site of an immunologically mediated sensitivity reaction to C. parvum. J. natn. Cancer Inst. *55:* 1403 (1975).
1531 Tyring, C. K.; Klimpel, G.; Brysk, M.; Gupta, V.; Stanton, G. J.; Fleischmann, R., Jr.; Baron, S.: Eradiation of cultured human melanoma cells by immune interferon and leukocytes. J. natn. Cancer Inst. *73:* 1067 (1984).

1532 Uchida, A.; Hoshino, T.: Reduction of suppressor cells in cancer patients treated with OK 432 immunotherapy. Int. J. Cancer 26: 401 (1980).

1533 Uhlenbruck, G.: The Thomson-Friedenreich (TF) receptor: an old history with new mystery. Immunol. Commun. 10: 251 (1981).

1534 Uhlenbruck, G.: Metastasenhemmung durch Lektin-Besetzung? Ärztl. Prax. 34: 1723 (1982).

1535 Uhlenbruck, G.; Paroe, G. I.; Bird, G. W. G.: On the specificity of lectins with a broad agglutination spectrum. II. Studies on the nature of the T-antigen and the specific receptors for the lectin or Arachis hypogoea (Ground Nut). Z. Immun-Forsch. 138: 423 (1969).

1536 Umiel, T.; Trainin, N.: Immunological enhancement of tumor growth by syngeneic thymus-derived lymphocytes. Transplantation 18: 244 (1974).

1537 Unanue, E. R.: The regulatory role of macrophages in antigenic stimulation. Adv. Immunol. 15: 95 (1972).

1538 Unanue, E. R.: Secretory function of mononuclear phagocytes. Am. J. Path. 83: 396 (1976).

1539 Unanue, E. R.: Cooperation between mononuclear phagocytes and lymphocytes in immunity. New Engl. J. Med. 303: 977 (1980).

1540 Unanue, E. R.: The regulatory role of macrophages in antigenic stimulation. II. Symbiotic relationship between lymphocytes and macrophages. Adv. Immunol. 31: 1 (1981).

1541 Unanue, E. R.; Askonas, B. A.: Persistence of imunogenicity of antigen after uptake by macrophages. J. exp. Med. 127: 915 (1968).

1542 Unanue, E. R.; Cerottini, J.-C.; Bedford, M.: Persistence of antigen on the surface of macrophages. Nature 222: 1193 (1969).

1543 Unanue, E. R.; Kiely, J.-M.: Synthesis and secretion of a mitogenic protein by macrophages: Description of a superinduction phenomenon. J. Immun. 119: 925 (1977).

1544 Unkeless, J.; Gordon, S.; Reich, F.: Secretion of plasminogen activator by stimulated macrophages. J. exp. Med. 139: 834 (1974).

1545 Urbanitz, D.; Büchner, T.; Pielken, H.; Van de Loo, J.: Immunotherapy in the treatment of acute myelogenous leukemia (AML): Rationale, results and future prospects. Klin. Wschr. 61: 947 (1983).

1546 Urbanitz, D.; Pielken, H.-J.; Koch, P.; Büchner, T.; Hiddemann, W.; Heinecke, A.; Wendt, F.; Maschmeier, G.; van de Loo, J.: Immunotherapy with allogeneic neuraminidase-treated blasts for maintenance in acute myelogenous leukemia (AML). Onkologie 8: 157 (1985).

1547 Uytdehaag, F.; Heijnen, C. J.; Pot, K. H.; Ballieux, R. E.: Antigen-specific human T cell factors. II. T cell suppressor factor: biologic properties. J. Immun. 126: 503 (1981).

1548 Vaage, J.: Non virus-associated antigens in virus-induced mouse mammary tumors. Cancer Res. 28: 2477 (1968).

1549 Vaage, J.; Agarwal, S.: Stimulation or inhibition of immune resistance against metastatic or local growth of a C3H mammary carcinoma. Cancer Res. 36: 1831 (1976).

1550 Vaessen, R. T. M. J.; Kreike, J.; Groot, G. S. P.: Protein transfer in nitro cellulose filters. A simple method for quantitation of single proteins in complex mixtures. FEBS Lett. 124: 193 (1981).

1551 Valentine, F.; Golomb, F.; Fazzini, E.: Effects of in vivo autologous lymphokines (LK) on injected and on noninjected metastatic nodules of human malignant melanoma. Int. J. Immunopharmacol. 4: 375 (1982).
1552 Van Heyningen, V.; Baron, L.; Brock, D. J. H.; Lawrie, S.: Monoclonal antibodies to human-fetoprotein: analysis of the behaviour of three different antibodies. J. immunol. Methods 50: 123 (1982).
1553 Vanky, F.; Stjernswärd, J.; Klein, G.; Nilsonne, V.: Serum-mediated inhibition of lymphocyte stimulation by autochthonous human tumors. J. natn. Cancer Inst. 47: 95 (1971).
1554 Vanky, F.; Argov, S.: Human tumour-lymphocyte interaction in vitro. VII. Blastogenesis and generation of cytotoxicity against autologous tumour biopsy cells are inhibited by interferon. Int. J. Cancer 26: 405 (1980).
1555 Vanky, F. T.; Argov, S. A.; Einhorn, S. A.; Klein, E.: Role of alloantigens in natural killing. Allogenic but not autologous tumor biopsy cells are sensitive for interferon-induced cytotoxicity of human blood lymphocytes. J. exp. Med. 151: 1151 (1980).
1556 Vánky, F.; Péterffy, Á.; Böök, K.; Willems, J.; Klein, E.; Klein, G.: Correlation between lymphocyte-mediated auto-tumor reactivities and the clinical course. II. Evaluation of 69 patients with lung carcinoma. Cancer Immunol. Immunother. 16: 17 (1983).
1557 Van Pel, A.; Georlette, M.; Boon, T.: Tumor cell variants obtained by mutagenesis of a Lewis lung carcinoma cell line: immune rejection by syngeneic mice. Proc. natn. Acad. Sci. USA 76: 5282 (1979).
1558 Van Voorhis, W. C.; Hair, L. S.; Steinman, R. M.; Kaplan, G.: Human dendritic cells: Enrichment and characterization from peripheral blood. J. exp. Med. 155: 1172 (1982).
1559 Vanwijck, R. R.; Goodrick, E. A.; Smith, H. G.; Goldweitz, J.; Wilson, R. E.: Stimulation or suppression of metastases with graded doses of tumor cells. Cancer Res. 31: 1559 (1971).
1560 Vogler, W. R.; Chan, Y. K.: Prolonging remission in myeloblastic leukaemia by tice-strain bacillus Calmette-Guerin. Lancet i: 128 (1974).
1561 Voisin, G. A.: Role of antibody classes in the regulatory facilitation reaction. Immunol. Rev. 49: 3 (1980).
1562 Von Boehmer, H.; Haas, W.; Jerne, N. K.: Major histocompatibility complex-linked immune-responsiveness is acquired by lymphocytes of low-responder mice differentiating in thymus of high-responder mice. Proc. natn. Acad. Sci. USA 75: 2439 (1978).
1563 Von Boehmer, H.; Turton, K.; Haas, W.: The role of the left end of the $H-2^b$ haplotype in the male-specific cytotoxic T cell response. Eur. J. Immunol. 9: 913 (1979).
1564 Von Dungern, E.: Über Immunität gegen Geschwülste. Münch. med. Wschr. 56: 1099 (1909).
1565 Von Leyden, V. E.; Blumenthal, F.: Vorläufige Mitteilungen über einige Ergebnisse der Krebsforschung auf der 1. medizinischen Klinik. Dt. med. Wschr. 28: 637 (1902).
1566 Vose, B. M.: Natural killers in human cancer: Activity of tumor-infiltrating and draining node lymphocytes; in Herberman, Natural cell-mediated immunity against tumors, p. 1081 (Academic Press, New York 1980).

1567 Wagner, H.; Hardt, C.; Heeg, K.; Pfizenmaier, K.; Stötter, H.; Röllinghoff, M.: The in vivo effects of interleukin 2 (TCGF). Immunobiology *161:* 139 (1982).

1568 Wahlin, B.; Perlmann, H.; Perlmann, P.: Analysis by a plaque assay of IgG- or IgM-dependent cytolytic lymphocytes in human blood. J. exp. Med. *144:* 1375 (1976).

1569 Waksman, B. H.: Adjuvants and immune regulation by lymphoid cells. Semin. Immunopathol. *2:* 5 (1979).

1570 Waldron, J. A., Jr.; Horn, R. G.; Rosenthal, A. S.: Antigen-induced proliferation of guinea pig lymphocyte in vitro: functional aspects of antigen handling by macrophages. J. Immun. *112:* 746 (1974).

1571 Wallack, M. K.: Specific immunotherapy with vaccinia oncolysates. Cancer Immunol. Immunother. *12:* 1 (1981).

1572 Wallack, M. K.; Steplewski, Z.; Koprowski, H.; Rosato, E.; George, J.; Hulhian, B.; Johnson, J.: A new approach in specific active immunotherapy. Cancer *39:* 560 (1977).

1573 Wang, M.; Halliday, W. J.: Immune responses of mice to iodoacetate-treated Ehrlich ascites tumour cells. Br. J. Cancer *21:* 346 (1967).

1574 Wang, B. S.; Onikul, S. R.; Mannick, J. A.: Prevention of death by metastasis by immune RNA. Science *202:* 59 (1978).

1575 Wang, T. J.; Miller, H. C.; Esselman, W. J.: Neuraminidase sensitivity of Thy-1 active glycoconjugates. Mol. Immunol. *17:* 1389 (1980).

1576 Wang, A. M.; Creasey, A. A.; Ladner, M. B.; Lin, L. S.; Strickler, J.; Van Arsdell, J. N.; Yamamoto, R.; Mark, D. F.: Molecular cloning of the complementary DNA for human tumor necrosis factor. Science *228:* 149 (1985).

1577 Warner, N. L.; Szenberg, A.; Burnet, F. M.: The immunological role of different lymphoid organs in the chicken. I. Dissociation of immunological responsiveness. Aust. J. exp. Biol. med. Sci. *40:* 373 (1962).

1578 Warner, N. L.; Tai, A. S.: Analysis of the differentiation lineage on NK cells. Fed. Proc. *40:* 2711 (1981).

1579 Warner, J. F.; Dennert, G.: Effects of a cloned cell line with NK activity on bone marrow transplants, tumour development and metastasis in vivo. Nature, Lond. *300:* 31 (1982).

1580 Weiden, P. L.; Storb, R.; Tsoi, M. S.; Deeg, H. J.; Graham, T. C.: Canine osteosarcoma: results of amputation with and without adjuvant immunotherapy. Cancer Immunol. Imunother. *5:* 181 (1978).

1581 Weiden, P. L.; Deeg, H. J.; Graham, T. C.; Storb, R.: Canine osteosarcoma: failure of intravenous or intralesional BCG as adjuvant immunotherapy. Cancer Immunol. Immunother. *11:* 69 (1981).

1582 Weidmann, E.; Sedlacek, H. H.; Lehmann, H. G.; Seiler, F. R.: Klinische Ergebnisse der Tumorimmuntherapie mit Neuraminidase behandelten Tumorzellen. In: Proceedings: 3. Arbeitsgespräch für Klinische Tumorimmunologie in der Gynäkologie, Bonn (1982).

1583 Weil, R.: Viral tumor antigens: A novel type of mammalian regulator protein. Biochim. biophys. Acta *516:* 301 (1978).

1584 Weinberg, J. B.; Chapman, H. A.; Hibbs, J. B.: Characterization of the effects of endotoxin on macrophage tumor cell killing. J. Immun. *121:* 72 (1978).

1585 Weinhold, K. J.; Miller, D. A.; Wheelock, E. F.: The tumor dormant state. Com-

parison of L5178Y cells used to establish dormancy with those that emerge after its termination. J. exp. Med. *149:* 745 (1979).
1586 Weiss, L.: Neuraminidase, sialic acids, and cell interactions. J. natn. Cancer Inst. *50:* 3 (1973).
1587 Weiss, D. W.: MER and other mycobacterial fractions in the immunotherapy of cancer. Med. Clins N. Am. *60:* 473 (1976).
1588 Weiss, D. W.: The questionable immunogenicity of certain neoplasms. Cancer Immunol. Immunother. *2:* 11 (1977).
1589 Welsh, R. M., Jr.: Mouse natural killer cells: Induction specificity and function. J. Immun. *121:* 1631 (1978).
1590 Werb, Z.; Gordon, S.: Secretion of a specific collagenase by stimulated macrophages. J. exp. Med. *142:* 346 (1975).
1591 Werb, Z.; Gordon, S.: Elastase secretion by stimulated macrophages chracterization and regulation. J. exp. Med. *142:* 361 (1975).
1592 Werner, G. H.: Immunopotentiating substances with antiviral activity. Pharmacol. Ther. *6:* 235 (1979).
1593 Whaley, K.: Biosynthesis of the complement components and the regulatory proteins of the alternative complement pathway by human peripheral blood monocytes. J. exp. Med. *151:* 501 (1980).
1594 Whitehead, J. S.; Kim, Y. S.: An inhibitor of lymphocyte proliferation produced by a human colonic adenocarcinoma cell line in culture. Cancer Res. *40:* 29 (1980).
1595 Whiteside, M. G.; Cauchi, M. N.; Paton, C.; Stone, J.: Chemoimmunotherapy for maintenance in acute myeloblastic leukemia. Cancer *38:* 1581 (1976).
1596 Whittaker, J. A.; Bailey-Wood, R.; Hutchins, S.: Active immunotherapy for the treatment of acute myelogenous leukaemia: Report of two controlled trials. Br. J. Haemat. *45:* 389 (1980).
1597 Wiener, E.; Levanon, D.: Macrophage cultures: an extracellular esterase. Science *159:* 217 (1968).
1598 Wikstrand, C. J.; Bigner, D. D.: Expression of human fetal brain antigens by human tumors of neuroectodermal origin as defined by monoclonal antibodies. Cancer Res. *42:* 267 (1982).
1599 Williams, T.; Granger, G.: Lymphocyte in vitro cytotoxicity: mechanism of human lymphotoxin-induced target cell destruction. Cell. Immunol. *6:* 171 (1973).
1600 Williams, R. M.; Dorf, M. E.; Benacerraf, B.: H-2 linked genetic control of resistance to histocompatibility tumors. Cancer Res. *35:* 1586 (1975).
1601 Williamson, B. D.; Carswell, E. A.; Rubin, B. Y.; Prendergast, J. S.; Old, L. J.: Human tumor necrosis factor produced by human B-cell lines: synergistic cytotoxic interaction with human interferon. Proc. natn. Acad. Sci. USA *80:* 5397 (1983).
1602 Weillmott, N.; Pimm, M. V.; Baldwin, R. W.: C. parvum treatment of transplanted rat tumours of spontaneous origin. Int. J. Cancer *24:* 323 (1979).
1603 Wilson, D. B.: Quantitative studies on the behavior of sensitized lymphocytes in vitro. I. Relationship of the degree of destruction of homologous target cells to the number of lymphocytes and to the time of contact in culture and consideration of the effects of isoimmune serum. J. exp. Med. *122:* 143 (1965).
1604 Wilson, R. E.; Sonis, S. T.; Godrick, E. A.: Neuraminidase as an adjunct in the treatment of residual systemic tumor with specific immune therapy. Behring Inst. Mitt. *55:* 334 (1974).

1605 Wilson, B. S.; Imai, K.; Natali, P. G.; Ferrone, S.: Distribution and molecular characterization of a cell surface and a cytoplasmic antigen detectable in human melanoma cells with monoclonal antibodies. Int. J. Cancer 28: 293 (1981).
1606 Wilson, B. S.; Kay, N. E.; Imai, K.; Ferrone, S.: Heterogeneity of human melanoma-associated antigens defined by monoclonal antibodies and conventional xenoantisera. Cancer Immunol. Immunother. 13: 69 (1982).
1607 Wing, E. J.; Gardner, I. D.; Ryning, F. W.; Remington, J. S.: Dissociation of effector functions in populations of activated macrophages. Nature, Lond. 268: 642 (1977).
1608 Winn, H. J.: Immune mechanisms in homotransplantation. I. The role of serum antibody and complement in the neutralization of lymphoma cells. J. Immun. 84: 530 (1960).
1609 Winn, H. J.: Immune mechanisms in homotransplantation. II. Quantitative assay of the immunological activity of lymphoid cells stimulated by tumor monografts. J. Immun. 86: 228 (1961).
1610 Wissler, R. W.; Flax, M. H.: Cytotoxic effects of antitumor sera. Ann. N. Y. Acad. Sci. 69: 773 (1957).
1611 Wolf, A.: The activity of cell-free tumour fractions in inducing immunity across a weak histocompatibility barrier. Transplantation 7: 49 (1969).
1612 Wolf, A.; Parry, D. M.; Barfoot, R. K.: Loss of weak antigenicity of lymphoma cells following treatment with difluoro-dinitrobenzene. Transplantation 10: 340 (1970).
1613 Wolfe, S. A.; Tracey, D. E.; Henney, C. S.: Induction of ‚natural killer' cells by BCG. Nature, Lond. 262: 584 (1976).
1614 Wolfe, S. A.; Tracey, D. E.; Henney, C. S.: BCG-induced murine effector cells. II. Characterization of natural killer cells in peritoneal exudate. J. Immun. 119: 1152 (1977).
1615 Wong, A.; Mankovitz, R.; Kennedy, J. C.: Immunosuppressive and immunostimulatory factors prodcued by malignant cells in vitro. Int. J. Cancer 13: 530 (1974).
1616 Wood, G. W.; Gillespie, G. Y.: Studies on the role of macrophages in regulation of growth and metastasis of murine chemically induced fibrosarcomas. Int. J. Cancer 16: 1022 (1975).
1617 Woodruff, M. F. A.: Cellular heterogeneity in tumours. Br. J. Cancer 47: 589 (1983).
1618 Woodruff, M. F. A.; Nolan, B.: Preliminary observations on treatment of advanced cancer by injection of allogeneic spleen cells. Lancet ii: 426 (1963).
1619 Woodruff, M. F.; Inchley, M. P.; Dunbar, N.: Further observations on the effect of C. parvum and anti-tumour globulin on syngeneically transplanted mouse tumours. Br. J. Cancer 26: 67 (1972).
1620 Woodward, J. H.; Daynes, R. A.: Cell-mediated immune response to syngeneic UV-induced tumors. IV. The presence of I-A and I-E subregion-coded antigens on the accessory cell required for the in vitro differentiation of cytotoxic T lymphocytes. J. Immun. 123: 1227 (1979).
1621 Woodward, J. G.; Fernandez, P. A.; Daynes, R. A.: Cell-mediated immune response to syngeneic UV-induced tumors. III. Requirement for an Ia+ macrophage in the in vitro differentiation of cytotoxic T lymphocytes. J. Immun. 122: 1196 (1979).
1622 World Health Organization: Immunological adjuvants. WHO Technical Report Series 595: 3 (1976).

1623 Wortzel, R. D.; Philipps, C.; Schreiber, H.: Multiple tumour-specific antigens expressed on a single tumour cell. Nature, Lond. *304:* 165 (1983).
1624 Wright, P. W.; Bernstein, I. D.: Serotherapy of malignant disease. Prog. exp. Tumor Res., vol. 25, p. 140 (Karger, Basel 1980).
1625 Wright, S. C.; Bonavida, B.: Selective lysis of NK-sensitive target cells by a soluble mediator released from murine spleen cells and human peripheral blood lymphocytes. J. Immun. *126:* 1516 (1981).
1626 Wu, R. L.; Kearney, R.: Specific tumor immunity induced with mitomycin C-treated syngeneic tumor cells (MCT). Effects of carrageenan and trypan blue on MCT-induced immunity in mice. J. natn. Cancer Inst. *64:* 81 (1980).
1627 Wunderlich, J. R.; Martin, W. J.; Fletcher, F.: Enhanced immunogenicity of syngeneic tumor cells coated with concanavalin A. Fed. Proc. *30:* 246 (1971).
1628 Wunderlich, M.; Schiessel, R.; Rainer, H.; Rauhs, R.; Kovats, E.; Schemper, M.; Dittrich, C.; Micksche, M.; Sedlacek, H. H.: Effect of adjuvant chemo- or immunotherapy on the prognosis of colorectal cancer operated for cure. Br. J. Surg. *72:* suppl p. 107 (1985).
1629 Yamamoto, H.: Isolation of immunogenic and suppressogenic glycoproteins from adenovirus type 12 hamster tumor cells. Microbiol. Immunol. *28:* 339 (1984).
1630 Yamamura, Y.; Yoshizaki, K.; Azuma, I.: Immunotherapy of human malignant melanoma with oil-attached BCG cell-wall skeleton. Gann *66:* 355 (1975).
1631 Yamashita, U.; Hamaoka, T.: The requirement of Ia-positive accessory cells for the induction of hapten-reactive cytotoxic T lymphocytes in vitro. J. Immun. *123:* 2637 (1979).
1632 Yamazaki, M.; Shinoda, H.; Suzuki, Y.; Mizumo, D.: Two-step mechanism of macrophage-mediated tumor lysis in vitro. Gann *67:* 741 (1976).
1633 Yamazaki, M.; Shinoda, H.; Hattori, R.; Mizuno, D.: Inhibition of macrophage-mediated cytolysis by lipoproteins from cell-free tumorous ascites. Gann *68:* 513 (1977).
1634 Yasuda, T.; Dancey, G. F.; Kinsky, S. C.: Immunogenicity of liposomal model membranes in mice: dependance on phospholipid composition. Proc. natn. Acad. Sci. USA *74:* 1234 (1977).
1635 Yeh, M. Y.; Hellstrom, I.; Brown, J. P.: Cell surface antigens of human melanoma identified by monoclonal antibody. Proc. natn. Acad. Sci. USA *76:* 2927 (1979).
1636 Yeh, M. Y. et al.: Clonal variation in expression of a human melanoma antigen defined by a monoclonal antibody. J. Immun. *126:* 1312 (1981).
1637 Yeh, M. Y.; Hellstrom, I.; Abe, K.; Hakomori, S.; Hellstrom, K. E.: A cell-surface antigen which is present in the ganglioside fraction and shared by human melanomas. Int. J. Cancer *29:* 269 (1982).
1638 Yin, H. L.; Bianco, C.; Cohn, Z. A.: The iodination and turnover of macrophage plasma membrane polypeptides; in van Furth, Mononuclear phagocytes: Functional aspects, part II, p. 25 (Nijhoff, The Hague 1980).
1639 Yoshida, M.; Miyoshi, I.; Hinuma, Y.: Isolation and characterization of retrovirus (ATLV) from cell lines of adult T-cell leukemia and its implication in the disease. Proc. natn. Acad. Sci. USA *79:* 2031 (1982).
1640 Young, W. W., Jr.; Hakomori, S. I.; Durdik, J. M.; Henney, C. S.: Identification of ganglio-N-tetraosylceramide as a new cell surface marker for murine natural killer (NK) cells. J. Immun. *124:* 199 (1980).

1641 Yu, D. T. Y.: Human lymphocyte receptor movement induced by sheep erythrocyte binding: effect of temperature and neuraminidase treatment. Cell. Immunol. *14:* 313 (1974).

1642 Yu, A.; Watts, H.; Jaffe, N.: Concomitant presence of tumor-specific cytotoxic and inhibitor lymphocytes in patients with osteogenic sarcoma. New Engl. J. Med. *297:* 121 (1977).

1643 Yuan, D.; Hendler, F. J.; Vitetta, E. S.: Characterization of a monoclonal antibody reactive with a subset of human breast tumors. J. natn. Cancer Inst. *68:* 719 (1982).

1644 Zagury, D.: Direct analysis of individual killer T cells. Susceptibility of target cells to lysis and secretion of hydrolytic enzymes by CTL; in Clark, Goldstein, Mechanisms of cell-mediated cytotoxicity, p. 149 (Plenum Press, New York 1982).

1645 Zarling, J. M.; Sosman, J.; Eskra, L.; Boren, E.; Horoszewicz, J. S.; Carter, W. A.: Enhancement of T cell cytotoxic responses by purified human fibroblast interferon. J. Immun. *121:* 2002 (1978).

1646 Zarling, J. M.; Eskra, L.; Borden, E. C.; Horoszewicz, J.; Carter, W. A.: Activation of human natural killer cells cytotoxic for human leukemia cells by purified interferon. J. Immun. *123:* 63 (1979).

1647 Zarling, J. M.; Kung, P. C.: Monoclonal antibodies which distinguish between human NK cells and cytotoxic T lymphocytes. Nature *288:* 394 (1980).

1648 Zarling, J. M. et al.: Phenotypes of human natural killer cell populations detected with monoclonal antibodies. J. Immun. *127:* 2575 (1981).

1649 Zbar, B.; Hunter, J. T.; Rapp, H. J.; Conti, G. F.: Immunotherapy of bilateral lymph node metastases in guinea pig by intralesional and paralesional injection of Mycobacterium bovis (BCG). J. natn. Cancer Inst. *60:* 1163 (1978).

1650 Zehngebot, L. M.; Alexander, M. A.; DuPont, G., IV.; Cines, D. B.; Mitchell, K.; Herlyn, M.: Functional consequence of variation in melanoma antigen expression. Cancer Immunol. Immunother. *16:* 30 (1983).

1651 Zeltzer, P. M.; Seeger, R. C.: Microassay using radioiodinated protein A from Staphylococcus aureus for antibody bound to cell surface antigens of adherent tumor cells. J. immunol. Methods *17:* 163 (1977).

1652 Zembala, M.; Asherson, G. L.: Contact sensitivity in the mouse. V. The role of macrophage cytophilic antibody in passive transfer and the effect of trypsin and anti-gamma globulin serum. Cell. Immunol. *1:* 276 (1970).

1653 Zembala, M.; Asherson, G. L.: Thymus derived cell suppression of contact sensitivity in the mouse. II. The role of soluble suppressor factor and its interaction with macrophages. Eur. J. Immunol. *4:* 799 (1974).

1654 Zettergren, J. G.; Luberoff, D. E.; Pretlow, T. G.: Separation of lymphocytes from disaggregated neoplasms by sedimentation in gradients of culture medium. J. Immun. *111:* 836 (1973).

1655 Zidek, Z.; Capková, J.; Boubelík, M.; Masek, K.: Opposite effects of the synthetic immunomodulator, muramyl dipeptide, on rejection of mouse skin allografts. Eur. J. Immunol. *13:* 859 (1983).

1656 Ziegler, H. K.; Henney, C. S.: Studies on the cytotoxic activity of human lymphocytes. II. Interactions between IgG and Fc receptors leading to inhibition of K cell function. J. Immun. *119:* 1010 (1977).

1657 Zinkernagel, R. M.: Thymus and self-MHC-restricted T cells. Behring Inst. Mitt. *70:* 118 (1982).

1658 Zinkernagel, R. M.; Doherty, P. C.: H-2 compatibility requirement for T-cell-mediated lysis of target cells infected with lymphocytic choriomeningitis virus. J. exp. Med. *141:* 1427 (1975).
1659 Zinkernagel, R. M.; Doherty, P. C.: MHC-restricted cytotoxic T cells: studies on the biological role of polymorphic major transplantation antigens determining T-cell restriction-specificity, function, and responsiveness. Adv. Immunol. *27:* 51 (1979).
1660 Zur Hausen, H.: Human genital cancer: Synergism between two virus infections or synergism between a virus infection and initiating events? Lancet *ii:* 1370 (1982).
1661 Zwisler, O.; Biel, H.: Über Elektrophorese in horizontalem Polyacrylamid-Gel. 3. Mitteilung: Die Kombination von Polyacrylamid-Gel-Elektrophorese und Immundiffusion. Behring Inst. Mitt. *46:* 129 (1966).